Field Guide to Fishes
of the Chesapeake Bay

Field Guide to Fishes of the Chesapeake Bay

Edward O. Murdy and John A. Musick

Illustrated by Val Kells

THE JOHNS HOPKINS UNIVERSITY PRESS   BALTIMORE

© 2013 The Johns Hopkins University Press
Illustrations © 2013 Val Kells
All rights reserved. Published 2013
Printed in China
9 8 7 6 5 4 3 2 1

The Johns Hopkins University Press
2715 North Charles Street
Baltimore, Maryland 21218-4363
www.press.jhu.edu

Library of Congress Cataloging-in-Publication Data

Murdy, Edward O.
   Field guide to fishes of the Chesapeake Bay / Edward O. Murdy and John A.
Musick ; illustrated by Val Kells.
      p.   cm.
   Includes index.
   ISBN 978-1-4214-0768-5 (pbk. : alk. paper)—ISBN 1-4214-0768-X (pbk. : alk.
paper)
   1. Fishes—Chesapeake Bay (Md. and Va.)—Identification. I. Musick, John A.
II. Title.
   QL628.C5M867 2013
   597.09163'47—dc23   2012023937

A catalog record for this book is available from the British Library.

*Special discounts are available for bulk purchases of this book. For more information,
please contact Special Sales at 410-516-6936 or specialsales@press.jhu.edu.*

The Johns Hopkins University Press uses environmentally friendly book
materials, including recycled text paper that is composed of at least 30 percent
post-consumer waste, whenever possible.

This book is dedicated to the memory of our close friend and colleague, Dr. John Edward Olney Sr., professor and chair at the Department of Fisheries Science at the Virginia Institute of Marine Science. John was an internationally recognized expert in fish systematics, ecology, and fisheries management. John directed studies of Chesapeake Bay fishes that spanned three decades and ranged from larval taxonomy and ecology to all aspects of population biology and beyond. He provided cogent advice on both commercial and recreational species to fisheries management agencies responsible for sustainable fisheries in the Chesapeake Bay and along the Atlantic Coast. His contributions to our knowledge of Chesapeake Bay fishes are immeasurable.—JAM, EOM, VAK

To my wife, Becky, who has always indulged my interest in fishes—EOM

To Bev, partner in all things—JAM

To my family: Andrew, Drew, and Dave Middleditch—VAK

CONTENTS

Appendices

ACKNOWLEDGMENTS

We began this project at the request of Vincent Burke, executive editor at the Johns Hopkins University Press. Vince believed a field guide would be a useful supplement to *Fishes of Chesapeake Bay*, published by the Smithsonian Institution Press in 1997. We are grateful to Vince for his initial suggestion and subsequent support (and prodding). Vince guided us through the preparation of this field guide with wise counsel and good humor.

It is difficult to list all the people who helped this effort reach fruition, but we must mention several individuals who provided especially important assistance. We thank Eric Durell and Paul Piavis of the Department of Natural Resources, State of Maryland, for sharing survey data for the Maryland portion of the Chesapeake Bay as well as local common names. Paul Gerdes, collection manager at the Virginia Institute of Marine Sciences (VIMS), provided access to the collection and use of the facilities. At the Smithsonian Institution, EOM thanks the following staff members of the Division of Fishes, National Museum of Natural History, for varied kinds of assistance: Jerry Finan, Kris Murphy, Sandra Raredon, and Jeff Williams. Especially helpful was Rich Vari, who provided work space and other courtesies to EOM. Mark Suskin of the National Science Foundation approved an independent research and development plan for EOM that enabled him to work on this book. We are grateful to Howard Weinberg (Chesapeake Bay Program) for preparing the original version of the art showing the seasonal salinities in the Chesapeake Bay and sending the image to Val Kells so she could modify it for our purposes. This is contribution #3162 from VIMS.

Field Guide to Fishes
of the Chesapeake Bay

Introduction

Watershed, History, and Hydrology of the Chesapeake Bay

The Chesapeake Bay watershed drains more than sixty-four thousand square miles and encompasses parts of Delaware, Maryland, New York, Pennsylvania, Virginia, and West Virginia, as well as the entire District of Columbia. The Chesapeake Bay is the largest of the 850 estuaries that bracket the United States, and it is fed by more than 150 rivers, creeks, and streams. The bay's three largest tributaries contribute more than 80% of the total freshwater input. The Susquehanna River is the largest tributary and, on average, contributes about half of the freshwater that enters the bay. The Potomac and the James Rivers are the next two largest tributaries and contribute 18% and 15% of the bay's freshwater, respectively. The bay mainstem is approximately 333 km (200 mi) in length and varies in width from 5.7 km (3.4 mi) near Aberdeen, Maryland, to 58 km (35 mi) near the mouth of the Potomac. The surface area of the bay is shared nearly equally by Maryland and Virginia, with the Virginia portion being slightly larger.

The Chesapeake Bay is a geologically young feature that had its beginnings near the end of the last ice age almost twenty thousand years ago. At that time, sea level was about 100 m (330 ft) lower than today because so much water was still trapped in glaciers and ice sheets. About eighteen thousand years ago, the climate began warming and the ice sheets began to melt. The ancient Susquehanna, carrying the outflow from glaciers far to the north, carved out the great channels in the Chesapeake Basin. As the glaciers receded, sea level slowly rose and progressed inland, burying the entire terrestrial biota beneath the coastal sea. With rising sea level, the Chesapeake Bay was formed and assumed its present shape about three thousand years ago. The Chesapeake Bay is still growing, albeit slowly, with water levels rising approximately 15 cm (6 in) per century. Anthropogenic climate change, however, may accelerate the rate of sea-level rise. Today, the deepest part of the bay is located off Bloody Point about 10 miles northwest of the mouth of

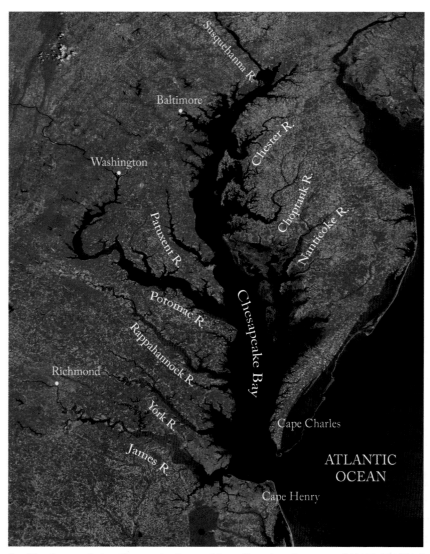

Satellite image of the Chesapeake Bay and its tributaries. Courtesy of NASA/Goddard Space Flight Center Scientific Visualization Studio.

the Choptank River and measures 53 m (174 ft) in depth. Actually a shallow water body, less than 10% of the area of the Chesapeake Bay is 18 m (60 ft) deep, and approximately 50% is less than 6 m (20 ft) deep.

As the sea invaded the freshwater tributaries of the Chesapeake Basin, brackish habitats were created: marshes developed on the intertidal lowlands, aquatic grass beds flourished in the shallow, flooded flatlands, and oysters formed acres of subtidal, hard-bottom reef habitat. Freshwater from the tributaries flows down the bay at the surface because it is less dense than saltwater from the ocean. At the same time, seawater has a net flow along the bottom, upstream into the bay and its tributaries. This incoming "salt wedge," which transports larval crabs and fishes and other organisms into the bay from the ocean, is a critical mechanism by which many species travel from the coastal areas where they have been spawned to the productive estuarine nursery areas, where they may grow rapidly in relatively protected shallow habitats, such as seagrass beds and marsh creeks. The eastern side of the bay is saltier than the western because of the Coriolis effect associated with rotation of the Earth and because of the inflow of many large freshwater tributaries on the bay's Western Shore (only about 7% of the bay's freshwater is provided by Eastern Shore tributaries). Salinity, which is expressed as grams of salt per liter of water, is denoted by the per-mille symbol, ‰. Typically, the bay's salinity is graded from near full seawater (32‰) at the mouth to freshwater (<0.5‰) at its northern extreme. Because of fluctuations in the amount of precipitation, the average salinity typically varies from month to month, season to season, and year to year. At any given location the salinity, temperature, turbidity, and other water features are subject to varying changes because of the tides.

The Chesapeake Bay has one of the most extreme annual temperature ranges known for the world's coastal ecosystems. Water temperatures in the bay may be as high as 29°C (84°F) in late summer and as low as 0°C (32°F) in late winter. Temperature differences, as well as the vertical salinity gradient, contribute to the vertical stratification of bay waters. Because warm water is less dense than cold water, the water column becomes increasingly stratified as surface waters begin to warm during the spring. A layered structure is established, especially in the deeper portions of the bay, with lighter, warmer, less salty water in the upper water column separated from heavier, colder, saltier water in the lower water column. Between these two layers there is a distinct boundary layer of sharp density change, the pycnocline. The transport of materials (oxygen, nutrients, and such) is reduced across this boundary layer.

An important result of stratification is that deep bottom waters become isolated from their oxygen source, the near-surface waters. In deep, dimly lit bottom waters, where photosynthesis is minimal, organic materials (mainly

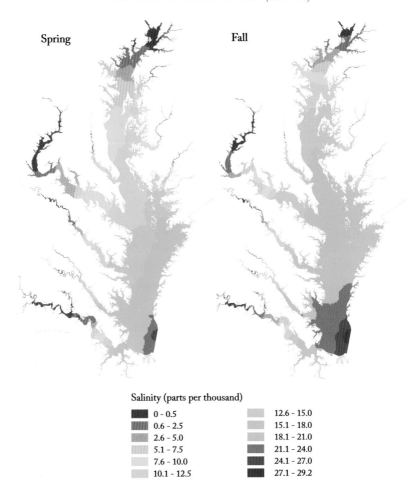

Salinity by season: (*Spring*) May 2005; (*Fall*) November 2005. Courtesy of Howard Weinberg and the Chesapeake Bay Program.

dead phytoplankton, but also zooplankton, fecal matter, and materials of human origin) sink from the productive waters above. In the bottom waters the organic matter is broken down by microbial action, with consumption of oxygen through microbial respiration. The greater the amount of organic matter introduced into the bottom waters, the greater the amount of oxygen consumed. During spring and summer, conditions of pronounced stratification, the low oxygen-carrying capacity of the water, and high organic production in the upper water column promote low levels of dissolved oxygen (hypoxia) in bottom waters. Throughout this period, oxygen levels may fall to near 0 mg/L, a level that causes extreme hypoxia in many aquatic organisms. Most fishes, mollusks, crustaceans, and other estuarine organisms show signs of distress when the dissolved oxygen falls below 2 mg/L, and some organisms are even more sensitive. Under extreme conditions these deep waters essentially become devoid of all but microbial life and are considered "dead zones," the first of which was reported in the Chesapeake Bay in the 1930s. The extent of summertime dead zones in bay waters has increased dramatically over the last 50 years and now typically includes most of the deeper parts of the bay's mainstem, from the Bay Bridge south. The southernmost part of the bay, which gets flushed more because of its proximity to the bay mouth, is less affected.

The cause of decreasing levels of dissolved oxygen is the excessive input of nutrients that overstimulate phytoplankton growth. The extreme growth of phytoplankton reduces water clarity, and the reduction of light penetration in turn reduces the growth of submerged rooted aquatic plants. The sources of the nutrients are numerous, but primary among them are runoff from agricultural and residential lands and municipal discharges. Although the environmental problems besetting the bay are numerous and varied, overenrichment is the most serious. It is estimated that, to greatly reduce or eliminate dead zones in the bay, we need a 40% reduction in nutrient loading from recent levels. This daunting task is made more difficult by population growth. During the last century, the human population of the Chesapeake Bay watershed grew from about two-and-a-half million to almost sixteen million. Experts predict that the population will increase to nearly twenty million by 2030. Only the most concerted public effort will stem the continued degradation of the bay as the human population of the watershed grows.

General Characteristics of Chesapeake Bay Fish Fauna

The flora and fauna of the Chesapeake Bay are particularly rich. More than thirty-six hundred species of plants and animals have been reported from its waters and tidal margins. The bay owes its variety to the incorporation of biotic elements from both the ocean habitat and the freshwater habitat, along with estuarine biota specifically adapted for life in waters of reduced salinity.

As the glaciers receded and the climate became warmer at the end of the Pleistocene, a diverse estuarine and marine fish fauna from warm-temperate and subtropical habitats to the south became established. Today these warmwater elements (such as killifishes, jacks, gobies, flounders, and drums) comprise the most abundant, diverse, and ecologically important groups of fishes in the Chesapeake Bay.

More than 350 species of fish can be found in the Chesapeake Bay and its tributaries. These can be divided into permanent residents and seasonal visitors. Depending on their feeding and spawning requirements, fishes may move between shallow and deep waters or between freshwater and saltwater throughout the year. Chesapeake Bay fishes can be placed in five categories: freshwater, estuarine, marine, anadromous, and catadromous. The first three categories are based on salinity regimes, whereas the last two categories consist of reproductively specialized fishes that migrate between freshwater and saltwater (or vice versa) in order to reproduce.

Freshwater fishes inhabit the tributaries of the bay, with many of these species descending into tidal freshwaters. Some of these fishes can tolerate brackish waters with salinities as high as 10‰. Salinity tolerance by freshwater fishes typically increases with decreasing temperature, and therefore their downstream penetration into brackish water is greatest in the winter. Only a few freshwater fishes ever enter the bay proper. In this field guide we treat only those freshwater fishes known to at least occasionally enter waters with salinities of 5‰ or greater.

Estuarine fishes typically inhabit tidal waters with salinities ranging from 0 to 30‰. Given this wide salinity tolerance, estuarine fishes can be found anywhere in the bay. There is a general tendency for estuarine fishes to penetrate farther up the tributaries and be closer to shore in the warmer months and to retreat to deeper water during the colder months.

Marine fishes typically spawn and spend much of their lives in coastal or oceanic waters with salinities greater than 30‰. Only a few marine species penetrate the bay north of the Potomac River. All the marine species migrate to and from the bay in response to the extreme seasonal changes in temperature there.

Anadromous fishes are those that migrate from ocean waters to freshwater to spawn. These fishes are often targeted by fisheries because the habit of anadromy ensures that large numbers will arrive in inland waters at a predictable time. Notable anadromous fishes in the bay include the shads, river herring, and striped bass. Upstream migration may extend to nontidal freshwaters. After spawning, the adults return downstream to the bay and eventually swim out to sea. A variety of fishes move from waters of higher salinity to those of lower salinity to spawn and can be categorized as semianadromous. These include fishes (such as white perch) that move from brackish water to freshwater to spawn as well as those (such as black drum) that migrate from oceanic waters to the slightly reduced salinities just inside the bay mouth for spawning.

Catadromous fishes display a spawning migration that takes them from freshwater to the ocean for spawning. True catadromy is rare in the northern hemisphere, and in the bay only the American eel can be so categorized.

Seasonal Fish Faunal Changes

The fish fauna of the Chesapeake Bay is dynamic because of the extreme seasonal temperature changes and the diversity of habitats within the bay. Fish diversity reaches a maximum in late summer and early autumn, when rarer tropical species may straggle into the bay to join warm-temperate and subtropical summer residents. With the onset of shorter day lengths in late summer, many marine species move toward the mouth of the bay. When the first of the cool northerly winds arrive in early autumn, most marine species begin their migration coastally to the south or to offshore waters or both. As large numbers of smaller drum-like fishes, menhaden, mullets, and other small fishes leave the bay and migrate down the coast, they are accompanied by predators of moderate size such as bluefish, weakfish, and young sandbar sharks.

As autumn progresses in the bay, cold-temperate species such as red, spotted, and silver hakes enter the lower bay to feed. But with the arrival of winter and cold temperatures ($\leq 4°C$, or $\leq 39°F$), even these cold-adapted species move back out onto the continental shelf, where water temperatures are more moderate. In midwinter, many of the more mobile estuarine resident species such as white perch, smaller striped bass, and young-of-the-year Atlantic croaker move into the deeper channels of the tributaries, where water temperatures are more stable. Of all the drum-like fishes found in the Chesapeake Bay, Atlantic croakers are the last to spawn in the fall, and their young are generally too small to migrate south with the spot, the weakfish, and other drum-like fishes. During warm winters ($\geq 5°C$, or $\geq 41°F$), young Atlantic croakers may have excellent survival, but in very cold winters ($\leq 4°C$, or $\leq 39°F$), most of the new year-class may die. Thus, Atlantic croaker year-class abundance in the Chesapeake Bay may fluctuate widely from year to year. The same is true of most marine species that use the bay as a nursery, although the mechanisms that determine year-class survival may differ among species.

During midwinter, the diversity and density of demersal (bottom) fishes in the Chesapeake Bay become quite low. However, winter flounder have been reported as common at times in the upper bay, and pelagic cold-temperate species such as Atlantic mackerel, Atlantic herring, and spiny dogfish often visit the lower bay in midwinter. By late winter and early spring (February–March), the anadromous alewife and American shad enter the bay and begin to ascend the tributaries on their spawning migrations. They are followed shortly by hickory shad, blueback herring, and striped bass. By late April some of the drum-like fishes and summer flounder have returned to the lower bay, and by late May most of the warm-temperate and subtropical summer residents have returned.

Conservation and Environmental Management of Chesapeake Bay Fishes

Trends in the commercial and recreational landings of fishes in the Chesapeake Bay are extremely changeable. This variability may be attributed to several factors. First, the boom and bust life-history strategies of most of our fishes lead to highly variable natural recruitment (addition of individuals to a population) from year to year. Second, the explosive growth of the human population in the Chesapeake Bay region and adjacent Mid-Atlantic Bight area has led to increased recreational and commercial fishing efforts, resulting in overexploitation of many fish species. In fact, important commercial species in the bay, such as American shad, river herring, American eel, summer flounder, weakfish, and yellow perch, have all experienced significant catch declines in comparison to the 1950s and 1960s. The specter of overfishing and the resultant collapse of fish stocks led to the creation of Fishery Management Plans by the Atlantic States Marine Fisheries Commission (ASMFC). As an extreme example, in 1998, the ASMFC closed the entire East Coast to Atlantic sturgeon fishing for the next 40 years, the longest fishing moratorium on record. In addition, the Chesapeake Bay Atlantic sturgeon population was listed as Endangered by the National Marine Fisheries Service in 2012. With the implementation of Fishery Management Plans, recovery of bay fish species has begun. One success story, the striped bass—once a severely overfished species along the entire Atlantic coast—was placed under a five-year moratorium in the 1980s. Striped bass stocks now appear to be fully recovered and the fishery was reopened, but with stringent minimum size limits, creel limits, a substantial closed season, and a commercial quota. Strict regulations have also been implemented for summer flounder, bluefish, weakfish, red drum, black drum, Spanish mackerel, and black sea bass as well as other species, and recovery has been achieved or is underway for all these fishes.

Because the bay's marine fishes are migratory, fishermen from Massachusetts to North Carolina may compete with Chesapeake Bay fishermen for the same stocks of fish, but at different times of the year, or at different times in the fish's life history. In addition, the Chesapeake Bay serves as an important nursery ground for several species—such as summer flounder, striped bass, bluefish, weakfish, and menhaden—that may migrate to other coastal areas, as far as Massachusetts to the north and North Carolina to the south, as they grow older. Thus, the environmental quality of the Chesapeake Bay region could impact survival, recruitment, and subsequent availability of fish to fisheries over a wide geographic area.

The Scope of This Field Guide

Organization

This field guide treats 211 fish species that are known to occur in the Chesapeake Bay as delimited below. These fishes include permanent residents, spawning migrants, and seasonal visitors. Of these 211 species, only 32 are year-round residents of the bay (table 1). The remainder enter the bay from either freshwater or the Atlantic Ocean for periods from days to months to feed, reproduce, or seek refuge.

In establishing the scope of this field guide, we have attempted to avoid arbitrary limits. Estuaries by definition are water bodies that grade from oceanic to freshwater. During times of flood, a number of freshwater fishes may be found in locations that we normally classify as estuarine. We have chosen to limit the inclusion of primary freshwater fishes to those that are commonly found in salinities of 5‰ or higher. The seaward limit of coverage is a line between Cape Henry and Cape Charles. We have included only those species taken west of this line and for which there is corroboration, normally a voucher specimen in the collections of the Virginia Institute of Marine Science or the U.S. National Museum of Natural History. The bay mouth is near the southern geographical limit of many northern species and the northern limit of many southern species; therefore, the recorded fish fauna includes many uncommon to rare transient marine fishes. Species as dissimilar in their primary habitats as the tropical to subtropical spotfin butterflyfish and the cold-temperate cunner are occasionally taken in the bay mouth. In the species accounts, we often refer to a species occurring in the upper, middle, and/or lower bay. These designations are associated with salinity levels and can be approximated as follows: Upper Bay—from the Susquehanna River to the Bay Bridge near the mouth of the Chester River; Middle Bay—from the Bay Bridge to the Potomac River; Lower Bay—the Virginia portion of the bay south of the Potomac River.

Year-to-year fluctuations in climate can cause wide variation in the abundance or occurrence of species, especially the less common. Very rare species

TABLE 1. RESIDENT FISHES IN THE CHESAPEAKE BAY

Common Name	Scientific Name	Common Name	Scientific Name
Bay anchovy	*Anchoa mitchilli*	Dusky pipefish	*Syngnathus floridae*
White catfish	*Ameiurus catus*	Northern pipefish	*Syngnathus fuscus*
Oyster toadfish	*Opsanus tau*	Naked goby	*Gobiosoma bosc*
Skilletfish	*Gobiesox strumosus*	Seaboard goby	*Gobiosoma ginsburgi*
Sheepshead minnow	*Cyprinodon variegatus*	Green goby	*Microgobius thalassinus*
Banded killifish	*Fundulus diaphanus*	Striped blenny	*Chasmodes bosquianus*
Mummichog	*Fundulus heteroclitus*	Feather blenny	*Hypsoblennius hentz*
Spotfin killifish	*Fundulus luciae*	Silver perch	*Bairdiella chrysoura*
Striped killifish	*Fundulus majalis*	Northern stargazer	*Astroscopus guttatus*
Bayou killifish	*Fundulus pulvereus*	White perch	*Morone americana*
Rainwater killifish	*Lucania parva*	Striped bass	*Morone saxatilis*
Rough silverside	*Membras martinica*	Yellow perch	*Perca flavescens*
Inland silverside	*Menidia beryllina*	Blackcheek tonguefish	*Symphurus plagiusa*
Atlantic silverside	*Menidia menidia*	Hogchoker	*Trinectes maculatus*
Fourspine stickleback	*Apeltes quadracus*	Smallmouth flounder	*Etropus microstomus*
Lined seahorse	*Hippocampus erectus*	Windowpane	*Scophthalmus aquosus*

These fishes are considered year-round residents of the Chesapeake Bay and complete all aspects of their natural history, including reproduction, in bay waters.

or those for which Chesapeake Bay records are in question are listed in table 2 and described in appendix 4. Undoubtedly, there are other marine species not included here that enter the bay on rare occasions but for which no substantiated records exist. If you find a fish that should fall within the scope of this field guide, but does not appear to be included, we urge you to make your find available to the Virginia Institute of Marine Science or one of the other marine institutes, laboratories, or universities around the bay.

Species accounts are provided for all fishes likely to be encountered in the bay as well as for some of the rarer ones. In each species account, information on the ecology, geographic distribution, and human interest for the fish in question is provided, along with descriptive features. Genera, species, and common names for species follow those of the American Fisheries Society. Fishes are presented in the text, tables, and appendices in an evolutionary sequence that is based on available research. Species within families are in alphabetical order. For each species account, a color illustration of the subject fish is provided.

How to Identify Fishes in the Bay

To ensure proper identification of Chesapeake Bay fishes, we recommend the following procedure:

TABLE 2. FISH SPECIES RARELY RECORDED FROM THE CHESAPEAKE BAY

Common name	Scientific name	Common name	Scientific name
Nurse shark	*Ginglymostoma cirratum*	Whitefin sharksucker	*Echeneis neucratoides*
Lemon shark	*Negaprion brevirostris*	Whalesucker	*Remora australis*
Tiger shark	*Galeocerdo cuvier*	Marlinsucker	*Remora osteochir*
Blacktip shark	*Carcharhinus limbatus*	Dolphinfish	*Coryphaena hippurus*
Atlantic torpedo	*Torpedo nobliana*	Lesser amberjack	*Seriola fasciata*
Spotted eagle ray	*Aetobatus narinari*	Cubera snapper	*Lutjanus cyanopterus*
Manta	*Manta birostris*	Atlantic tripletail	*Lobotes surinamensis*
Bonefish	*Albula vulpes*	Tidewater mojarra	*Eucinostomus harengulus*
Speckled worm eel	*Myrophis punctatus*	Slender mojarra	*Eucinostomus jonesii*
Hardhead catfish	*Ariopsis felis*	Flagfin mojarra	*Eucinostomus melanopterus*
Rainbow trout	*Oncorhynchus mykiss*	Tomtate	*Haemulon aurolineatum*
Coho salmon	*Oncorhynchus kisutch*	White grunt	*Haemulon plumierii*
Atlantic salmon	*Salmo salar*	Atlantic threadfin	*Polydactylus octonemus*
Brook trout	*Salvelinus fontinalis*	Barbu	*Polydactylus virginicus*
Brown trout	*Salmo trutta*	Red goatfish	*Mullus auratus*
Snakefish	*Trachinocephalus myops*	Dwarf goatfish	*Upeneus parvus*
White hake	*Urophycis tenuis*	Yellow chub	*Kyphosus incisor*
Sargassumfish	*Histrio histrio*	Bermuda chub	*Kyphosus sectatrix*
Longnose batfish	*Ogcocephalus corniger*	Blue parrotfish	*Scarus coeruleus*
Atlantic agujon	*Tylosurus acus*	Crested blenny	*Hypleurochilus geminatus*
Atlantic flyingfish	*Cheilopogon melanurus*	Fat sleeper	*Dormitator maculatus*
Squirrelfish	*Holocentrus adscensionis*	Clown goby	*Microgobius gulosus*
Bluespotted cornetfish	*Fistularia tabacaria*	Code goby	*Gobiosoma robustum*
Leopard searobin	*Prionotus scitulus*	Cero	*Scomberomorus regalis*
Bighead searobin	*Prionotus tribulus*	Bluefin tuna	*Thunnus thynnus*
Sea raven	*Hemitripterus americanus*	Atlantic halibut	*Hippoglossus hippoglossus*
Flying gurnard	*Dactylopterus volitans*	Yellowtail flounder	*Limanda ferruginea*
Goliath grouper	*Epinephelus itajara*	Spotted whiff	*Citharichthys macrops*
Rock sea bass	*Centropristis philadelphica*	Scrawled cowfish	*Acanthostracion quadricornis*
White crappie	*Pomoxis annularis*	Checkered puffer	*Sphoeroides testudineus*
Walleye	*Sander vitreus*	Porcupinefish	*Diodon hystrix*
Bigeye	*Priacanthus arenatus*	Ocean sunfish	*Mola mola*
Short bigeye	*Pristigenys alta*		

The fish species listed here have been recorded from the Chesapeake Bay, but in many instances their occurrence has been recorded only once. A specimen validating such an occurrence may or may not exist. Listed species are not expected to be found in the bay in any abundance or with any regularity.

1. Familiarize yourself with the sections "Morphology" and "Basic Counts and Measurements" below and with the glossary of selected technical terms at the end of the field guide, because this information is critical to understanding the technical language used in the keys. Words not defined in the glossary can be found in a standard dictionary.
2. To identify a particular fish, work your way through the keys, beginning with the most general key (appendix 1, the key to the orders and fami-

lies) and then the keys to species within families (see appendix 3). Both sets of keys will refer the reader to the appropriate page(s) in the guide. See the section "How to Use the Keys," below.

3. Compare the fish with the text description and illustration of the species in the species account. If they correspond, your identification is probably correct.

How to Use the Keys

The keys in the back of the book provide a rather simple method for identifying a fish by eliminating, through a series of alternate choices, all groups of fishes (orders, families) except the one in question. When the possibilities have been narrowed down, the key will direct you to a page number where species accounts for a given family begin.

Each key consists of consecutively numbered couplets. Each couplet includes a pair of choices labeled "a" and "b." For example, begin with couplet 1 of the key to orders, and read the choices (be sure to read both choices before reaching a conclusion). Select the "a" or "b" choice that best describes your fish, and proceed as indicated by the notation at the end of the choice. If the notation is a number, continue in the same key to the couplet with that number, and make a choice as before. If the notation is the name of an order or family, then proceed to the indicated page in the text for that group and continue your identification process in the species accounts. Or, you may continue to appendix 3, where keys to species within families are provided. If the comparison of your fish matches the description in a species account, your fish is correctly identified. Your identification can be verified by referring to the illustration accompanying the species account. An additional tip: if neither choice within a couplet seems to match your specimen, or if the species account or illustration does not correspond to your fish, work backward through your previous couplet choices to ensure that you have correctly read each couplet and properly interpreted the characters in your specimen. If this process returns you to the same impasse, then follow each of the two choices in turn; usually one of them will quickly be revealed as the wrong path.

Morphology

To identify fishes, it is necessary to know something about their structure, especially the parts used in classification. The following general morphological terms refer to all animals:

Anterior: in front of; the front end of the body or structure
Posterior: behind; the back end of the body or structure
Dorsal: toward, near, or pertaining to the back or upper surface
Ventral: toward, near, or pertaining to the underpart or lower surface
Lateral: toward, near, or pertaining to the side
Medial: toward, near, or pertaining to the middle

The technical terms and principal measurements most commonly used in identification are illustrated in the figures.

Fishes have both paired and unpaired (median) fins. In sharks, skates, and rays, the fins are covered by thick skin such that the skeletal supports are not visible without dissection. The skeletal supports in the fins of bony fishes, however, are easily visible and may be present as hard pointed spines, flexible segmented rays, or both. The number of spines and/or rays in a given fin is frequently a useful diagnostic character. The pectoral and pelvic fins, when present, are paired. The pectoral fins are usually located laterally near the gill openings and the pelvic fins along the belly. Variations in the length and shape of these fins are useful characters in identification, as is the placement of the pelvic fins. The position of the pelvic fins is termed *abdominal* when they are inserted near the anus, *thoracic* when inserted near or directly ventral to the pectoral fins, and *jugular* when inserted anterior to the pectoral fins. The unpaired fins of fishes consist of the dorsal, anal, and caudal fins. The dorsal fin extends along the midline of the back and may be divided into several parts. A singular adipose fin or a series of finlets may be present posterior to the dorsal fin in some fishes. The adipose fin is fleshy and lacks spines or rays, whereas finlets are supported by a single soft ray. The anal fin is located along the ventral midline just posterior to the anus. The tail usually terminates in a caudal fin. There are many variations in the shape of the caudal fin; based on the arrangement of the internal bony support, the caudal fin of most fishes may be categorized as either heterocercal or homocercal. In the heterocercal tail, the vertebral column extends into the upper portion of the fin and the fin is asymmetrical, usually with the upper lobe much larger than

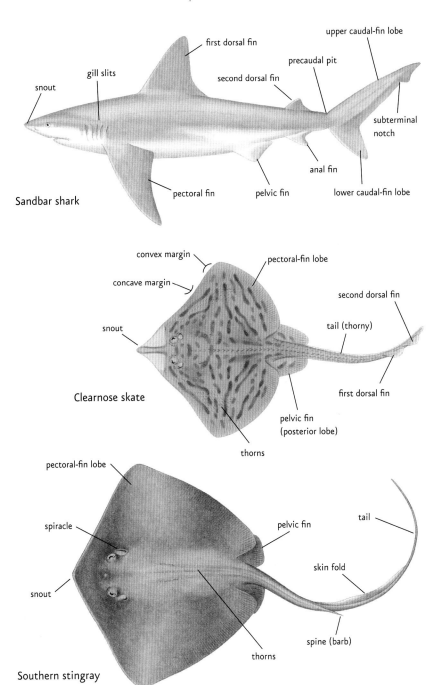

Sandbar shark

first dorsal fin

upper caudal-fin lobe

precaudal pit

second dorsal fin

gill slits

snout

subterminal notch

pectoral fin

pelvic fin

anal fin

lower caudal-fin lobe

Clearnose skate

convex margin

pectoral-fin lobe

concave margin

second dorsal fin

snout

tail (thorny)

first dorsal fin

pelvic fin (posterior lobe)

thorns

Southern stingray

pectoral-fin lobe

spiracle

pelvic fin

tail

snout

skin fold

spine (barb)

thorns

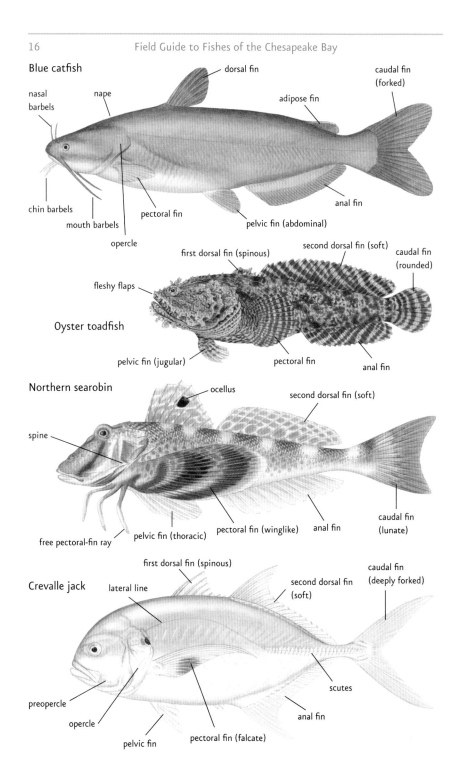

Blue catfish

dorsal fin

caudal fin (forked)

nasal barbels

nape

adipose fin

chin barbels

mouth barbels

opercle

pectoral fin

pelvic fin (abdominal)

anal fin

Oyster toadfish

first dorsal fin (spinous)

second dorsal fin (soft)

caudal fin (rounded)

fleshy flaps

pelvic fin (jugular)

pectoral fin

anal fin

Northern searobin

ocellus

second dorsal fin (soft)

spine

free pectoral-fin ray

pelvic fin (thoracic)

pectoral fin (winglike)

anal fin

caudal fin (lunate)

Crevalle jack

first dorsal fin (spinous)

lateral line

second dorsal fin (soft)

caudal fin (deeply forked)

preopercle

opercle

pelvic fin

pectoral fin (falcate)

anal fin

scutes

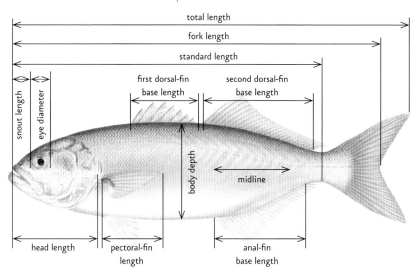

total length

fork length

standard length

snout length

eye diameter

first dorsal-fin base length

second dorsal-fin base length

body depth

midline

head length

pectoral-fin length

anal-fin base length

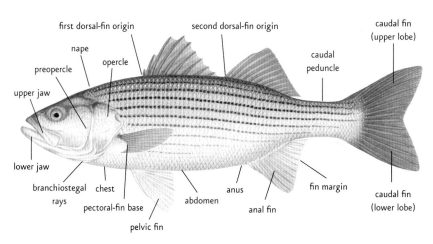

first dorsal-fin origin

second dorsal-fin origin

caudal fin (upper lobe)

nape

opercle

caudal peduncle

preopercle

upper jaw

lower jaw

branchiostegal rays

chest

pectoral-fin base

pelvic fin

abdomen

anus

anal fin

fin margin

caudal fin (lower lobe)

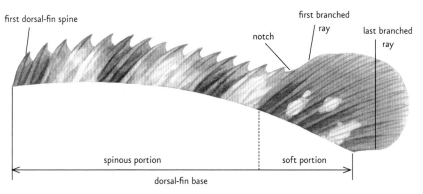

first dorsal-fin spine

first branched ray

last branched ray

notch

spinous portion

soft portion

dorsal-fin base

the lower. In sharks and sturgeons the caudal fin is strongly heterocercal. The abbreviate heterocercal (hemicercal) tail found in the bowfin and gars is less easily distinguished. Homocercal caudal fins have a variety of shapes but are typically symmetrical or nearly so, and the vertebral column does not extend into the tail.

The scales of bony fishes serve as an important tool in identification. Their presence or absence, type, and number along a given line are frequently utilized. The types of scales differentiated in these keys are ganoid, cycloid, and ctenoid. Ganoid scales are hard, thick, rhomboid or diamond-shaped, and barely overlapping. Cycloid scales are smooth, thin, rounded, and overlapping. Ctenoid scales are similar to cycloid scales, but the exposed portion bears tiny spines called ctenii. Ctenoid scales usually feel rough to the touch. Some bony fishes have an axillary scale, or axillary process, at the base of the pectoral and pelvic fins.

The head of a fish includes the gill region and corresponds to the head, neck, and throat of higher vertebrates. Many diagnostic characters are found in the head region. The snout is that portion of the head projecting forward from the anterior rim of the eye. It contains the nares (nostrils), a pair of blind pits that function in odor detection. Each naris usually has two openings, but in some fishes only one aperture is present. The upper jaw is ventral to the snout and in bony fishes includes several paired bones. The lower jaw, or mandible, consists of several bones. In some bony fishes a prominent bone, the gular plate, is present ventrally between the arms of the lower jaw. Some fishes may have fleshy, thread-like structures called barbels around the mouth and snout regions of the head.

The gill area offers another important region for differentiation. In sharks, skates, and rays each gill chamber has a separate opening to the outside, whereas the gills of bony fishes are usually enclosed in a chamber covered by a bony flap called the opercle, or operculum. The gill chamber is located under the operculum and contains the gills. Each set of gills consists of a pair of bony arches (pharyngeal arches) that support a double row of red gill filaments on their outer edge and a row of knobby, hairy, or finger-like structures called gill rakers on their inner edge. Gill rakers range in shape from knob-like bumps to filamentous hairs. The number of gill rakers and their shape and size are useful in the identification of many fishes.

Fishes have an external set of sensory structures collectively known as the lateral line system. The most obvious part of this system is a series of pores extending in a line along the sides of the trunk and tail. The presence or absence, as well as the configuration, of the lateral line system is helpful in identification.

Basic Counts and Measurements

The number of fin spines or rays is frequently used as a diagnostic character. For the purpose of this field guide, spines are unpaired structures without segmentation. They are usually stiff but may be rudimentary or flexible. Rays are usually branched, flexible, and segmented. When a fin contains both spines and rays, the spines are always anterior to the rays. In the dorsal and anal fins, the last two rays are usually close together and appear to come from a single origin. These two rays are conventionally counted as a single ray.

The most common scale count used in this text is the number of scales along the lateral line or along an imaginary line in the position that would normally be occupied by a typical lateral line. The count originates with the scale touching the shoulder girdle and ends at the base of the caudal fin. The base of the caudal fin is determined by the presence of a crease, which is clearly visible when the tail is bent to either side. Lateral line scales behind that crease are not counted, and if a scale lies directly over the crease, it is not counted if the middle of the scale is posterior to the crease.

The most common measurements called for in the text are standard length, total length, fork length, head length, and body depth. The type of length measurement that is used to describe a fish often depends on the particular characteristics of the fish in question. Standard length (SL) is the greatest distance in a straight line from the tip of the snout to the base of the caudal fin. Total length (TL) is the greatest distance in a straight line from the tip of the snout to the posteriormost tip of the caudal fin. Fork length (FL) is the distance from the tip of the snout to the midpoint of the caudal fin margin. Head length (HL) is the greatest distance in a straight line from the tip of the snout to the posteriormost point of the opercular membrane. Body depth is the greatest vertical distance in a straight line exclusive of fins or any fleshy or scaly structure associated with fin bases.

Other terms used in the text and keys may be found in the glossary or in a standard dictionary. A bibliography and additional resources are available online at www.chesapeakefishes.com and at www.press.jhu.edu.

Species Accounts

Sea lamprey - *Petromyzon marinus* Linnaeus, 1758

KEY FEATURES: Adults with eel-like, cartilaginous body; mouth circular with concentric rows of horny teeth; seven pore-like gill openings posterior to each eye; dorsal fin posterior and separated into two lobes; anal fin a small posterior fold; pectoral and pelvic fins absent. COLOR: Adults, brownish to yellowish gray with dark mottling on the back. Parasitic juveniles, dark gray on back with light belly. SIZE: Common to 84 cm (34 in) SL; 1 m (39 in) maximum. RANGE: North Atlantic coasts of Europe and North America, extending from Labrador to Florida in the western Atlantic. Introduced land-locked populations occur in the upper Great Lakes.

Sea lamprey *Petromyzon marinus*

HABITAT AND HABITS: Anadromous, ascending larger freshwater tributaries to spawn in the spring. After hatching, larvae burrow into silty substrates in backwaters and sluggish tributaries where they may remain for four to five years before metamorphosing into juveniles that descend into estuarine and coastal waters. There they become parasitic on larger fishes, attaching to their skin and rasping away flesh and blood with their horny teeth and sucker-like mouths. Larger sea lampreys are most often found on the continental shelf. Sea lampreys may remain in the marine environment for two years before maturing and returning to freshwater to spawn. OCCURRENCE IN THE CHESAPEAKE BAY: Sea lampreys are found in all larger freshwater tributaries as well as the main stem of the Chesapeake Bay.

REPRODUCTION: Sea lampreys ascend freshwater in Maryland and Virginia from March to June to locate suitable spawning grounds. Spawning occurs over gravel or cobble where the males excavate a nest that may be 0.75–1.0 m (29–39 in) in diameter. Females may deposit more than 250,000 eggs, and after spawning the adults die.

FOOD HABITS: After metamorphosis, juveniles and adults are parasitic, mostly on larger bony fishes but also on sharks and marine mammals.

IMPORTANCE: Although probably present in all major drainages of the Chesapeake Bay during their spring spawning period, sea lampreys are not present in sufficient numbers to be a major destructive force, even though they have been found attached to menhaden, bluefish, and weakfish.

Spiny dogfish - *Squalus acanthias* Linnaeus, 1758

KEY FEATURES: Slender, torpedo-shaped shark with spines in front of both dorsal fins; small keel on each side of base of tail; anal fin absent. COLOR: Dark gray with light spots above, lighter below. SIZE: Maximum adult size about 90 cm (2.9 ft) TL in males, and at least 1.1 m (3.6 ft) TL in females. RANGE: Circumglobal in cold-temperate seas except in the North Pacific. In the western North Atlantic, spiny dogfish occur from Newfoundland to the Gulf of Maine in summer and as far south as the coast of Georgia in winter.

Spiny dogfish *Squalus acanthias*

HABITAT AND HABITS: Spiny dogfish form large sexually segregated schools and make strong seasonal migrations, spending the summer mostly from Georges Bank north and migrating south in autumn to overwinter in large numbers, mostly off Virginia and North Carolina. In spring, dogfish return to their New England and Canadian summering areas. Spiny dogfish pup along the edge of the continental shelf from Georges Bank to Virginia. OCCURRENCE IN THE CHESAPEAKE BAY: Spiny dogfish visit the lower Chesapeake Bay from late autumn into spring. They are most commonly encountered at the bay mouth.

REPRODUCTION: Male spiny dogfish mature at six years of age. Female spiny dogfish mature at about 12 years of age and have a gestation period of 18–24 months, after which they may produce as many as 15 pups (average 6 or 7). These are small (20–33 cm, or 8–13 in, TL).

FOOD HABITS: Spiny dogfish may feed anywhere in the water column from the surface to the bottom. They consume a diversity of prey, ranging from comb jellyfish, squids, mackerel, and herring to a wide array of benthic fishes, shrimps, crabs, and even sea cucumbers.

IMPORTANCE: Spiny dogfish are harvested throughout their range by gillnets, trawls, bottom set lines, and traps. The history of spiny dogfish fisheries in most areas is one of rapid development, large harvests, and then stock collapse. Despite the apparent abundance of spiny dogfish stocks, they are vulnerable to rapid overfishing because of their late maturity, low fecundity, and two-year reproductive cycle. In the western North Atlantic, spiny dogfish were heavily overfished during the 1990s in response to European markets that preferred large (female) sharks. As a consequence, the most fecund members of the population were removed. As the fishery progressed, mostly males and immature females remained, and several years of recruitment failure resulted. Currently, the stock has been rebuilt and is being managed under a stringent plan. Spiny dogfish historically have been the target of derision by both recreational and commercial fishers because of their habit of appearing in large schools that sometimes resulted in lost bait, caused damage to fishing gear, and/or required labor to remove from the gear.

Bull shark - *Carcharhinus leucas* (Müller and Henle, 1839)

KEY FEATURES: Large rounded triangular first dorsal fin placed just at or behind pectoral fin; ridge absent between first and second dorsal fins. COLOR: Gray on back, white ventrally; fin tips dusky, particularly in young. SIZE: Maximum adult size 3.5 m (11.4 ft) TL, common to about 2.8 m (9 ft) TL. RANGE: Circumglobal in tropical and subtropical waters.

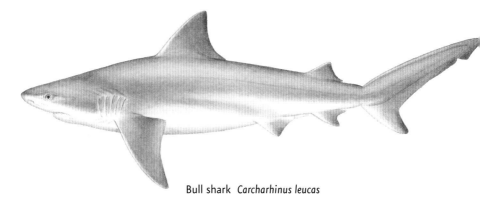

Bull shark *Carcharhinus leucas*

HABITAT AND HABITS: Bull sharks are coastal and will ascend rivers into freshwater as far as 4000 km (2400 mi) from the sea. They are among the most euryhaline of all sharks, commonly foraging in estuaries and coastal rivers. Bull sharks are resident in tropical and subtropical areas but may migrate into warm-temperate waters in summer. OCCURRENCE IN THE CHESAPEAKE BAY: Bull sharks are rare to occasional summer visitors to the Chesapeake Bay that will penetrate at least as far as the Patuxent River. Two individuals, both measuring almost 2.5 m (8.3 ft) TL, were caught by anglers near the mouth of the Potomac River in September 2010. With climate change and global warming, bull sharks can be expected to occur here more frequently in the future.

REPRODUCTION: Bull sharks produce from 1–13 free-swimming pups, supported during embryonic development by a placenta. Gestation is about nine months. Pupping occurs in brackish mangrove estuaries and river mouths in the spring. Bull sharks mature at about six years.

FOOD HABITS: Bull sharks eat a wide variety of bony fishes, skates, rays, and other sharks. When older, they consume larger prey such as sea turtles, marine mammals, and sharks. They are among a few species of sharks that actively hunt, attack, and dismember large prey. They may also scavenge.

IMPORTANCE: Bull sharks are responsible for more shark attacks on humans in the tropics than any other shark, with the possible exception of tiger sharks. Although there have been no recorded shark-related injuries from any species within the Chesapeake Bay, a large (3-m, or 10-ft) bull shark was implicated in the fatal attack that occurred in 2001 on a 10-year-old boy along the Virginia Beach oceanfront, less than 10 miles from the bay mouth.

Dusky shark - *Carcharhinus obscurus* (Lesueur, 1818)

KEY FEATURES: Short first dorsal fin (shorter than either the sandbar or bull shark) located behind pectoral fin; low ridge on middle of back between first and second dorsal fins. COLOR: Gray on back, with brassy overtones in life along sides, white below; fin tips dusky. SIZE: Largest individuals attain lengths of 3.5–4.0 m (about

Dusky shark *Carcharhinus obscurus*

11.5–13 ft) TL; more commonly, adults are in the range of 2.8–3.2 m (about 9–10 ft) TL. RANGE: Circumglobal in subtropical and warm-temperate seas.

HABITAT AND HABITS: Dusky sharks are coastal, occurring from shallow surf zone habitats to the edge of the continental shelf. In the western North Atlantic they undergo seasonal migrations, north in spring into the Mid-Atlantic Bight, and offshore and south in winter as far as the Gulf of Mexico and the Yucatán Peninsula. The shoals on the seaside of the Eastern Shore of Virginia serve as pupping and nursery grounds in summer. Juvenile dusky sharks overwinter south of Cape Hatteras off the North Carolina coast. OCCURRENCE IN THE CHESAPEAKE BAY: Dusky sharks sometimes enter the Chesapeake Bay mouth, but their occurrence further into the bay is limited by low salinity.

REPRODUCTION: Dusky sharks produce 3–14 pups that receive placental support during a gestation period that may last 18 months or longer. Females enter a resting stage for about one year after giving birth, resulting in a three-year reproductive cycle. The newborn pups may be as long as 1.0 m (3.3 ft) TL. Adults mature at 19–21 years of age. Such a low reproductive rate and late maturity result in one of the slowest population increase rates known among sharks.

FOOD HABITS: Dusky sharks consume a wide variety of bony fishes, skates, rays, and other sharks as well as squids, octopuses, and crustaceans.

IMPORTANCE: Dusky sharks are not targeted by either commercial or recreational fisheries in the Chesapeake Bay.

Sandbar shark - *Carcharhinus plumbeus* (Nardo, 1827)

KEY FEATURES: Very large triangular first dorsal fin placed over broad pectoral fins; dermal ridge on midline of back between dorsal fins; precaudal pit at base of tail. COLOR: Gray above, with brassy highlights along sides in life, white below. SIZE: Commonly to 2 m (6.5 ft) TL, maximum 2.4 m (7.8 ft) TL. RANGE: Circumglobal in warm-temperate to tropical coastal areas. Range in the western North Atlantic from the Gulf of Maine to the Yucatán in the Gulf of Mexico.

HABITAT AND HABITS: Sandbar sharks undertake seasonal migrations into more temperate waters in summer, returning to subtropical areas in winter. Island populations such as the one in Hawaii are residential. Sandbar sharks use estuaries as pupping and nursery grounds in regions where estuarine habitats are available. In the absence

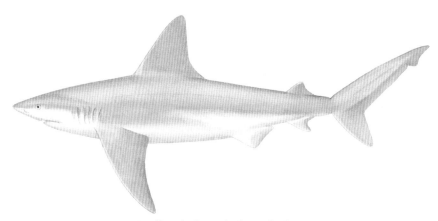

Sandbar shark *Carcharhinus plumbeus*

of available estuarine environments, they tend to use coastal reef habitats. OCCUR-
RENCE IN THE CHESAPEAKE BAY: Sandbar sharks are abundant summer residents in
the lower Chesapeake Bay, which serves as the principal pupping and nursery ground
for the northwest Atlantic population. Females give birth in late May and June and
then migrate offshore and north along the coast. Newborn sharks remain in the nursery ground, which is defined by the 20‰ salinity line. Thus, these sharks are mostly
absent from lower-salinity areas of the bay, and the nursery area expands during dry
summers and contracts toward the bay mouth in rainy years. As day length shortens
and bay water temperatures drop, the young sharks leave the bay and migrate south
of Cape Hatteras to coastal wintering areas near the Gulf Stream off North Carolina.
In spring as surface waters warm to about 18°C (64°F), juvenile sandbar sharks return
north to the Chesapeake Bay and other nearshore areas in the Mid-Atlantic Bight. As
they grow older, their fall migrations become longer, extending to wintering areas off
Florida and in the Gulf of Mexico. Males mostly remain south of Cape Hatteras and/
or offshore after about age eight. Females continue to return in summer north of Cape
Hatteras until they mature at about age 13. Thereafter females return to pup every
other summer, remaining in the south in alternate years while building up nutritional
reserves to support another pregnancy.

REPRODUCTION: Females give birth to live young after about a nine-month pregnancy. Developing embryos are supported by a placenta. Nine to ten young are born every
other year.

FOOD HABITS: Young sandbar sharks feed heavily on crustaceans such as blue crabs
and mantis shrimp but transition to a predominantly fish diet with age. Adults prey
mostly on bottom fishes such as croakers, small sharks, and skates but also eat some
pelagic species, such as menhaden.

IMPORTANCE: Sandbar sharks have been the most important large sharks in the directed commercial fishery along the Atlantic Coast and in the recreational fishery from New Jersey southward. Because they have slow growth, late maturity, and a small number of young, the species has been easily overfished, and between 1980 and 1992 the northwest Atlantic population declined by about 80%. Fishery management measures put in place during the 1990s stopped this decline, but recovery has been elusive. More stringent management measures were implemented by the National Marine Fisheries Service, and the Atlantic Marine Fisheries Commission established protection for the species in 2008.

Atlantic sharpnose shark - *Rhizoprionodon terraenovae*
(Richardson, 1836)

KEY FEATURES: Snout long and slender, longer than width of mouth; first dorsal fin above rear of pectoral fin; small second dorsal fin begins over middle of anal fin. COLOR: Gray above and white below; adults with numerous pale round spots; pectoral fin with pale trailing edge. SIZE: Maximum adult size at least 1.1 m (3.6 ft) TL. RANGE: Atlantic sharpnose sharks are found in warm-temperate to tropical waters of the western North Atlantic. These sharks are resident in the Gulf of Mexico, but in summer, they can stray as far north as the Bay of Fundy. They can also be found along the coast of Brazil in the western South Atlantic.

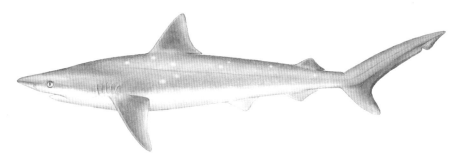

Atlantic sharpnose shark *Rhizoprionodon terraenovae*

HABITAT AND HABITS: Atlantic sharpnose sharks are demersal and abundant inshore summer visitors at least as far north as Maryland seaside waters. Although primarily a coastal species, these sharks have been found in offshore waters at depths to at least 280 m (910 ft). Juveniles often occur in the surf zone off sandy beaches. The population north of Cape Hatteras is dominated by adult males. In autumn, Atlantic sharpnose sharks migrate south and overwinter offshore in the South Atlantic Bight.

Atlantic sharpnose sharks may form large sexually segregated schools during migration. OCCURRENCE IN THE CHESAPEAKE BAY: Atlantic sharpnose sharks are rare to occasional visitors to the lower Chesapeake Bay from their more common coastal habitats.

REPRODUCTION: Atlantic sharpnose sharks mature at two or three years. Females produce 3–7 young that are supported during a 10-month gestation by placental nutrition. Pupping takes place in estuarine lagoons in the South Atlantic Bight and Gulf of Mexico during the spring. Mating occurs in spring after parturition.

FOOD HABITS: Atlantic sharpnose sharks feed on a wide variety of invertebrates and small fishes.

IMPORTANCE: Atlantic sharpnose sharks are the most numerous sharks landed in the directed small coastal commercial shark fishery, although they have lower value than any of the large coastal species. Despite these sources of fishery mortality, the Atlantic sharpnose shark population remains robust, probably because of their early maturity and annual reproduction.

Scalloped hammerhead - *Sphyrna lewini*
(Griffith and Smith, 1834)

Smooth hammerhead - *Sphyrna zygaena*
(Linnaeus, 1758)

KEY FEATURES: These species differ in the front margin of the head; scalloped hammerheads have an indentation centrally that smooth hammerheads do not. In addition, in scalloped hammerheads, the free rear tip of the second dorsal fin almost reaches the base of the tail, and the base of the second dorsal fin is shorter than the base of the anal fin, while in smooth hammerheads the free rear tip of the second dorsal fin falls far short of the tail base and the base of the second dorsal fin is about the same length as the base of the anal fin. COLOR: Scalloped hammerheads are yellowish brown above, while smooth hammerheads are dark gray to charcoal above. Both species are white below. SIZE: Both of these hammerheads may reach 4 m (13 ft) TL, but 3 m (9.8 ft) is more typical. RANGE: Scalloped hammerheads are circumtropical and subtropical, migrating into warm-temperate waters in summer. Smooth hammerheads are primarily subtropical and warm-temperate and are absent from the tropics in most areas. In the western North Atlantic, scalloped hammerheads occur from New York south into the tropics. Smooth hammerheads may range as far north as Nova Scotia in summer and occur south to Florida and the Antilles.

HABITAT AND HABITS: Large schools of scalloped hammerheads occur in winter in the Florida straits and migrate north in the Gulf Stream in spring, dispersing coastally in the Mid-Atlantic Bight in summer. In autumn they migrate south following their finfish prey even into the surf zone. Smooth hammerheads appear to be much less common in the Mid-Atlantic Bight and less is known of their habits, other than that they migrate into the region in summer and move south and offshore in winter. OCCURRENCE IN THE CHESAPEAKE BAY: Both species may occur at the bay mouth in summer.

REPRODUCTION: Both of these hammerheads give birth to live young and have large litters (scalloped, 15–31; smooth, 23–50). Pupping in scalloped hammerheads takes place inshore in late spring and summer, and the pups use estuarine and shallow coastal waters south of Cape Hatteras to Florida and the Gulf of Mexico as nurseries. Little is known of pupping and nursery areas for smooth hammerheads in the western North Atlantic.

FOOD HABITS: Both prey on a wide variety of bony fishes and squids. Larger individuals also feed on small sharks and stingrays.

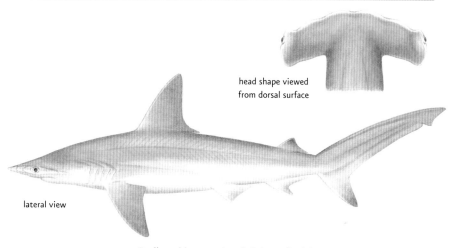

head shape viewed
from dorsal surface

lateral view

Scalloped hammerhead *Sphyrna lewini*

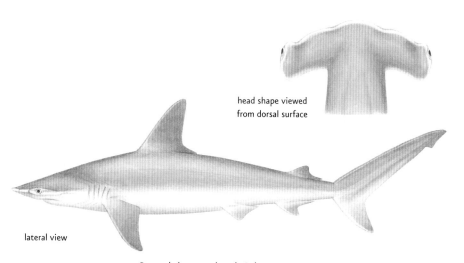

head shape viewed
from dorsal surface

lateral view

Smooth hammerhead *Sphyrna zygaena*

IMPORTANCE: In the western North Atlantic, scalloped hammerheads have undergone large declines over the last few decades due to overfishing. Population data are not available for smooth hammerheads, but it is not unreasonable to conclude that they have suffered the same sorts of declines.

Bonnethead - *Sphyrna tiburo* (Linnaeus, 1758)

KEY FEATURES: Spade-shaped head, with lateral expansions of head relatively short. **COLOR:** Gray to gray brown above, light below. **SIZE:** Maximum adult size 1.5 m (4.9 ft) TL, typically about 1.3 m (4.2 ft) TL. **RANGE:** In the western Atlantic, bonnetheads occur regularly in summer as far north as North Carolina (occasionally to southern New England) and throughout the southeastern United States, the Gulf of Mexico, and Central America to southern Brazil. Bonnetheads also occur in the eastern Pacific from southern California to Ecuador.

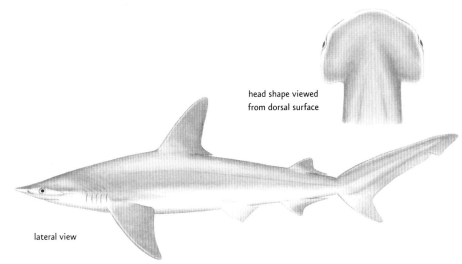

head shape viewed
from dorsal surface

lateral view

Bonnethead *Sphyrna tiburo*

HABITAT AND HABITS: Bonnetheads are a shallow inshore species found along the coast, from the surf zone to depths of 80 m (260 ft), in estuaries and channels, and on reefs and in seagrass beds. They spend the nighttime hours on shallow grass flats searching for nocturnally active invertebrate prey and move into deeper water during the day. Bonnetheads migrate north in the summer and south in the autumn and winter and usually occur in small (fewer than 15 individuals) schools. During migrations schools of hundreds or perhaps thousands may form. Sexual segregation is common. **OCCURRENCE IN THE CHESAPEAKE BAY:** Bonnetheads are occasional summer visitors to the lower Chesapeake Bay, particularly in and near Lynnhaven Inlet near the bay mouth.

REPRODUCTION: Bonnetheads are livebearers and produce 4–16 pups after a 4-month gestation. Pups are 35–40 cm (13–16 in) at birth and pupping occurs in shallow water in late summer and early fall. Bonnetheads take three years to mature. Females apparently produce litters every year.

FOOD HABITS: Bonnetheads consume mostly crustaceans, including crabs, mantis shrimp, and other shrimps. They have molar-like teeth in the back of their jaws that are particularly well-suited for crushing hard-shelled prey. Bonnetheads also feed on mollusks, octopuses, and small fishes.

IMPORTANCE: Bonnetheads are taken in all manner of inshore fisheries throughout their range and are eaten fresh, dried, or smoked. In the United States, bonnetheads are taken mostly as unwanted bycatch. They are the second most abundant small coastal shark (sharpnose sharks are first) in both the commercial and recreational fisheries of the United States.

Smooth dogfish - *Mustelus canis* (Mitchill, 1815)

KEY FEATURES: Teeth small, arranged as tiles in the jaws; first dorsal fin just behind pectoral fin, and second dorsal fin smaller than first; both dorsal fins with rounded apices; anal-fin origin at midpoint of second dorsal fin. COLOR: Uniformly grayish dorsally, with pale belly; can change its color with change in substrate (one of only a few sharks able to do so). SIZE: Maximum adult size 1.5 m (4.9 ft) TL. RANGE: Warm-temperate and subtropical waters of the western North Atlantic from the Gulf of Maine (occasionally) to the Gulf of Mexico and Antilles and in the western South Atlantic from southern Brazil to Argentina.

Smooth dogfish *Mustelus canis*

HABITAT AND HABITS: Smooth dogfish are demersal and coastal and migrate inshore seasonally into the Mid-Atlantic Bight in the spring. Adults can be found in summer from New Jersey to Massachusetts, where they typically inhabit waters less than 18 m

(59 ft) deep over mud or sand bottoms. In autumn, smooth dogfish migrate offshore as far as the edge of the continental shelf and south of Cape Hatteras. Smooth dogfish swim in packs or schools; this habit may be the basis for the use of the term *dogfish*. OCCURRENCE IN THE CHESAPEAKE BAY: Common to abundant seasonal visitors in summer and fall to the lower Chesapeake Bay, extending as far north as near the mouth of the Nanticoke River. Females pup at the mouth of the Chesapeake Bay from late April to early June. Juveniles use the lower Chesapeake Bay as a summer nursery.

REPRODUCTION: Smooth dogfish are placental and give birth to live young. Females produce 4–20 young each year after attaining maturity at about 4 years of age.

FOOD HABITS: Smooth dogfish are active swimmers and bottom feeders, preying primarily on large crustaceans such as crabs; they will also eat small fishes, squids, and bottom-dwelling invertebrates.

IMPORTANCE: Smooth dogfish are targeted in gillnet fisheries in North Carolina and Virginia from late April to early May. They are landed in other gillnet fisheries mostly as bycatch along the northeast U.S. coast. Thus far the smooth dogfish population has handled modest fishing pressure well, primarily because it has a higher population increase rate than most other sharks.

Basking shark - *Cetorhinus maximus* (Gunnerus, 1765)

KEY FEATURES: Five very long gill slits, almost meeting dorsally, long bristle-like gill rakers; caudal peduncle with strong lateral keels; caudal fin nearly symmetrical. COLOR: Dark gray above, lighter below. SIZE: The second largest sharks (whale

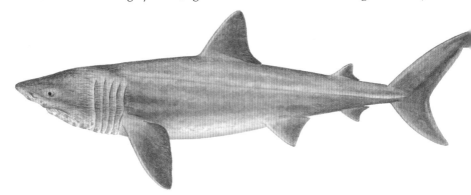

Basking shark *Cetorhinus maximus*

sharks are the largest), basking sharks routinely attain a length of 10 m (33 ft) TL or longer (14 m, or 45 ft). RANGE: Cold-temperate waters in both hemispheres.

HABITAT AND HABITS: Basking sharks are typically observed near the surface with the mouth wide open as they swim along collecting planktonic food. These coldwater pelagic sharks spend the summers in the Gulf of Maine to Newfoundland and migrate south in late autumn, overwintering from Virginia to Georgia. Some individuals may move into deep water at the edge of the continental shelf. Recent satellite tracking revealed that some basking sharks migrate as far south as the Caribbean in winter. There they occur at great depths, where water temperatures are cold and zooplankton may be concentrated. Basking sharks often occur in small groups (2–4 individuals), particularly in areas where plankton is plentiful, but aggregations containing as many as 100 individuals have been observed. OCCURRENCE IN THE CHESAPEAKE BAY: Basking sharks are rare visitors to the lower Chesapeake Bay during colder months, particularly in the early spring. Their occurrence is probably limited by the lower salinities in the bay.

REPRODUCTION: Basking sharks usually have 2 large (1.5–1.8 m, or 4.9–5.9 ft, TL) pups that are maternally supported during their embryonic development by consuming large numbers of yolk-rich unfertilized ova. Gestation time is unknown but may be at least two years.

FOOD HABITS: Basking sharks are filter feeders that strain zooplankton such as copepods, small shrimps, and fish eggs as well as small fishes from seawater that passes over their gill rakers. As much as 2000 gallons of seawater per hour are strained by a basking shark's gill rakers.

IMPORTANCE: Historically, basking sharks have been taken in harpoon and net fisheries throughout their range. The flesh has been consumed fresh and dried and salted, and the fins, hide, and carcass have also been utilized. Basking sharks have a very low reproductive rate and consequently have declined throughout their range from overfishing. They are protected along the Atlantic coast of the United States by federal fisheries regulations and are listed as Vulnerable on the International Union for the Conservation of Nature (IUCN) Red List.

Sand tiger - *Carcharias taurus* Rafinesque, 1810

KEY FEATURES: Teeth in upper and lower jaws similar, narrow, spike-like, recurved, with small cusp on each side of main tooth; two dorsal fins of equal size placed far back on body. COLOR: Tan to gray, with darker mottling. SIZE: Maximum to 3.3 m (10 ft) TL. RANGE: Circumglobal in temperate and subtropical (occasionally tropical) seas, except the eastern Pacific.

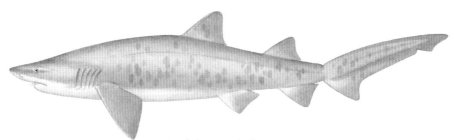

Sand tiger *Carcharias taurus*

HABITAT AND HABITS: Shallow coastal and estuarine habitats mostly less than 15 m (49 ft) deep, occasionally as deep as 190 m (618 ft). Sand tigers are the only known shark that regulate their buoyancy by swallowing air and descending to slowly cruise close to the bottom. Sand tigers prefer areas with structure, such as shipwrecks, and make seasonal migrations to higher latitudes in spring and lower latitudes in autumn. OCCURRENCE IN THE CHESAPEAKE BAY: Once common in the lower Chesapeake Bay, sand tigers now occur in smaller numbers, with maximum abundance in May, early June, and September. They migrate south of Cape Hatteras in the fall, overwintering as far south as Florida.

REPRODUCTION: Sand tiger females take 10 years to mature, while males take 7. Mating occurs on the coast between southern Florida and Cape Hatteras during the early spring. Gestation takes 9–12 months. Sand tigers, as with most other sharks, have paired uteri, but as fertilized eggs in each uterus begin to develop, the largest embryo on each side devours its siblings. In addition, the mother continues to produce unfertilized eggs that are consumed by the two remaining embryos for much of the rest of the pregnancy. The result is two very large (1-m, or 39-in) pups. Females enter a yearlong resting phase after pupping, resulting in a two-year reproductive cycle. Pupping takes place south of Cape Hatteras in spring, and the young sharks migrate north with older juveniles, adult males, and nonpregnant females to spend the summer mostly from Delaware to Massachusetts Bay.

FOOD HABITS: Sand tigers consume a wide variety of mostly demersal sharks, skates, rays, and bony fishes as well as some pelagic species such as menhaden. In the lower Chesapeake Bay, they are the principal predator of young sandbar sharks, which are abundant there.

IMPORTANCE: Sand tigers were once the target of an ardent recreational shark fishery in Virginia and elsewhere along the Atlantic Coast. Because they are often found in shallow water, they were the only large shark species consistently available to shore and pier anglers. In addition, sand tigers were harvested in the commercial bottom longline fishery for large coastal sharks. Consequently, sand tigers and several other large coastal species underwent declines in the 1980s and early 1990s. Despite ever more stringent fishery regulations, and complete protection in 1996, the sand tiger population has been slow to recover, probably because of their very limited pup production.

Atlantic angel shark - *Squatina dumeril* Lesueur, 1818

KEY FEATURES: Body flat, much wider than high (like a ray), but with distinct pectoral fins not attached to head; eyes dorsal; mouth broad and trap-like, teeth small and sharp; two small dorsal fins located close to tail; anal fin absent. COLOR: Tan to brown to gray above, with scattered dark spots; belly pale; abdomen, throat, and ventral fins with reddish spots. SIZE: Maximum adult size about 1.8 m (5.9 ft) TL, commonly from 90 to 120 cm (3–4 ft) TL. RANGE: Warm-temperate to tropical waters in the western North Atlantic. Found from Massachusetts to Florida and in the northern Gulf of Mexico, Jamaica, and the north coast of South America. Apparently absent from Central America.

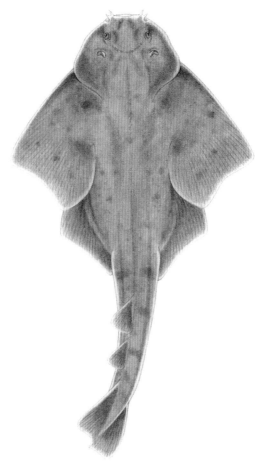

Atlantic angel shark *Squatina dumeril*

HABITAT AND HABITS: The Atlantic angel shark is coastal and benthic, preferring to lie on sandy or muddy bottoms where it may partially bury itself to ambush passing prey. Angel sharks migrate inshore and north into the Mid-Atlantic Bight in spring and offshore and south in autumn. They may occur from the surf zone to the shelf edge (180 m, or 585 ft) depending on the season but occur deeper in the southern part of the range (>1000 m, or 3280 ft). OCCURRENCE IN THE CHESAPEAKE BAY: Atlantic angel sharks are occasional summer or autumn visitors to the lower Chesapeake Bay. They rarely occur in the upper bay.

REPRODUCTION: Atlantic angel sharks produce 16–25 pups. The length of gestation and age at maturity are unknown. Pupping has been reported to occur on the inner continental shelf (20–30 m, or 65–98 ft).

FOOD HABITS: Atlantic angel sharks eat a wide variety of small demersal bony fishes, mollusks, and crustaceans.

IMPORTANCE: Atlantic angel sharks are not targeted in any recreational or commercial fishery. However, the species is often taken as unwanted bycatch in a variety of bottom trawl fisheries. Although the species is not dangerous in the water, on deck it will snap at people and should be handled with care. Because of its propensity to bite, this fish is also known as the "sand devil."

Smalltooth sawfish - *Pristis pectinata* Latham, 1794

KEY FEATURES: Shark-like appearance, but with gill slits located ventrally, as in skates and rays; long saw-like rostrum in front of head, with 24–32 teeth on each side of saw; large pectoral fins attached to head. **COLOR:** Body uniformly dusky; grayish brown above; white or grayish white below. **SIZE:** Maximum adult size 7.6 m

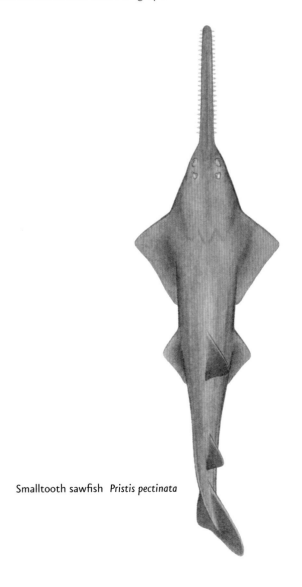

Smalltooth sawfish *Pristis pectinata*

(24.7 ft) TL, commonly reaching 5.5 m (17.9 ft) TL. RANGE: Circumtropical waters except in the eastern Pacific. In the western North Atlantic, reported from New York (as a stray) to Florida, the Bahamas, Gulf of Mexico, and Caribbean to Brazil. In the United States, recent records of this species are scarce, except for South Florida.

HABITAT AND HABITS: Smalltooth sawfish are primarily a shallow-water coastal species that can be found in brackish water as often as full seawater. They occur in estuaries, seagrass meadows, and mangrove channels and ascend coastal rivers far up into freshwater areas. OCCURRENCE IN THE CHESAPEAKE BAY: Historically, smalltooth sawfish were rare to occasional summer visitors to the lower Chesapeake Bay.

REPRODUCTION: Females produce 15–20 pups after about a 1-year gestation period. Pupping takes place in shallow brackish habitats during summer. In the United States, pupping occurs in the Everglades and adjacent areas in South Florida.

FOOD HABITS: Smalltooth sawfish use their "saw" to slash at and disable or impale small schooling fishes such as mullet and herring, after which the prey are recovered from the bottom and consumed. Smalltooth sawfish also stir up bottom sediments with their "saw" to expose benthic prey such as shrimps and crabs.

IMPORTANCE: Because of their tooth-studded rostrum, smalltooth sawfish have been vulnerable to entanglement in virtually all types of nets. In addition, the propensity of this species to readily take cut fish and other baits have made it vulnerable to line fisheries. The result has been a range contraction in federal waters of about 90%, with the last remnants of the population surviving in South Florida. Because of this severe population decline, smalltooth sawfish have been classified as Endangered under the Endangered Species Act.

Southern stingray - *Dasyatis americana*
Hildebrand and Schroeder, 1928

KEY FEATURES: Tips of disk abruptly angled; tail with broad fin-fold underneath about same width as tail, small ridge on top of tail. **COLOR:** Dark gray to brown on back, white below. **SIZE:** Maximum 2 m (6.5 ft) disk width (DW). **RANGE:** In the western Atlantic from Cape Cod to Florida, coastally throughout the Gulf of Mexico and Caribbean Sea, and southward to Argentina. The original description of the species is based on specimens from Crisfield, Maryland.

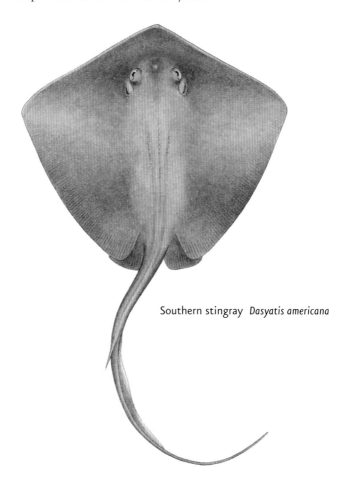

Southern stingray *Dasyatis americana*

HABITAT AND HABITS: Southern stingrays make seasonal migrations into warm-temperate latitudes in summer and return to warmer climes for the winter. They lie partially covered with sediment on sandy or muddy bottoms or slowly swim along with fins flapping just over the bottom in search of prey. Southern stingrays occur at depths from 1–53 m (3–172 ft) or greater. OCCURRENCE IN THE CHESAPEAKE BAY: Summer visitors to the Chesapeake Bay that are rarely encountered in the upper bay, extending as far north as the mouth of the Choptank River, and occasionally reported from the lower bay.

REPRODUCTION: Southern stingrays are viviparous, and the developing embryos are nourished by uterine "milk" secretions. Southern stingray males mature at 51 cm (20 in) DW and females at 75–80 cm (30–32 in) DW. Gestation has been estimated at 4–11 months, and pups are born in the summer. The number of pups is somewhat correlated with the size of the female and can range from 2 to 10 (4 is average).

FOOD HABITS: While slowly swimming near the bottom, southern stingrays locate food items using electroreceptors in combination with smell and touch. They eat a wide variety of prey, including small fishes, mantis shrimps and other crustaceans, worms, and mollusks. Southern stingrays possess powerful grinding teeth that enable them to crush even the toughest shells.

IMPORTANCE: There are no directed fisheries in the Chesapeake Bay for southern stingrays, but they are taken as unwanted bycatch. Southern stingrays must be handled with great care because the venomous spine can inflict extremely painful if not life-threatening wounds. Bathers should shuffle their feet in the water to avoid stepping on a partially buried stingray.

Roughtail stingray - *Dasyatis centroura* (Mitchill, 1815)

KEY FEATURES: Tips of disk abruptly angled; tail with narrow fin-fold underneath much narrower than width of tail, no ridge on top of tail; distinctive thorn-like bucklers on back of disk and on tail. COLOR: Dark gray to brown on back, white below. SIZE: Maximum 2.2 m (6.8 ft) DW. These are one of the largest stingrays. RANGE: Warm-temperate waters from Cape Cod to northeast Gulf of Mexico in the western North Atlantic. Southern Brazil to Argentina in the western South Atlantic. In the eastern Atlantic, Bay of Biscay to Angola and the Mediterranean Sea.

Roughtail stingray *Dasyatis centroura*

HABITAT AND HABITS: Roughtail stingrays make seasonal migrations onshore and into warm-temperate latitudes in summer and return offshore and to warmer climes for the winter but are absent from the tropics. They lie partially covered with sediment on sandy or muddy bottoms or slowly swim along with fins flapping just over the bottom in search of prey. Roughtail stingrays occur on the continental shelf from the surf zone out to depths of 274 m (890 ft). OCCURRENCE IN THE CHESAPEAKE BAY: Because roughtail stingrays are a warm-water, demersal species that undergoes summer migrations to higher latitudes, they occasionally enter the Chesapeake Bay.

REPRODUCTION: Roughtail stingrays are viviparous, and the developing embryos are nourished by uterine "milk" secretions. Roughtail stingray males mature at 1.3–1.5 m (4.2–4.9 ft) DW and females at 1.4–1.6 m (4.6–5.2 ft) DW. Gestation has been estimated at four months but may be much longer. Two to six pups are born in autumn.

FOOD HABITS: Roughtail stingrays eat a wide variety of prey, including fishes, squids, crustaceans, and worms.

IMPORTANCE: There are no directed fisheries in the Chesapeake Bay for roughtail stingrays, but they are taken as unwanted bycatch in all manner of fisheries. Roughtail stingrays must be handled with great care because the venomous spine can inflict extremely painful if not life-threatening wounds.

Atlantic stingray - *Dasyatis sabina* (Lesueur, 1824)

KEY FEATURES: Distance from eye to tip of snout considerably longer than distance between spiracles; front margin of disk concave; disk gently rounded, broadly angled. **COLOR:** Brown to yellowish brown on back, white below; fin-folds on tail brown to yellow. **SIZE:** Maximum 61 cm (2.4 ft) DW but typically 32–37 cm (1.3–1.5 ft) DW. This species is one of the smaller stingrays. **RANGE:** Subtropical waters in the western Atlantic from the Chesapeake Bay to Florida and the Gulf of Mexico to the Yucatán.

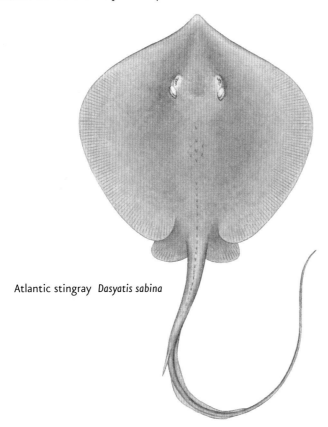

Atlantic stingray *Dasyatis sabina*

HABITAT AND HABITS: Atlantic stingrays occur in shallow inshore areas on sand or silty substrates, usually less than 25 m (81 ft) deep. Atlantic stingrays make seasonal migrations into warm-temperate latitudes in summer and return to warmer climes for the winter. They lie partially covered with sediment on sandy or muddy bottoms. Atlantic stingrays prefer shallow water but may migrate out onto the continental shelf to avoid cold temperatures in winter. In addition, they are the most tolerant of all our stingrays of fresh and brackish water occurring far up rivers. Freshwater populations occur in Florida. OCCURRENCE IN THE CHESAPEAKE BAY: Atlantic stingrays are a seasonal visitor to the lower Chesapeake Bay during the summer and fall that are usually not found farther north in the bay than the York River.

REPRODUCTION: Atlantic stingrays are viviparous, and the developing embryos are nourished by uterine "milk" secretions. Atlantic stingray males mature at 20 cm (8 in) DW and females at 24 cm (9 in) DW. Gestation has been estimated at four to five months, and pups are born in summer. The number of pups ranges from one to four.

FOOD HABITS: Atlantic stingrays locate their prey using electroreceptors and consume small creatures that live on or in the bottom, including burrowing anemones, small crustaceans, worms, mollusks, and brittle stars.

IMPORTANCE: There are no directed fisheries in the Chesapeake Bay for Atlantic stingrays, but they are taken as unwanted bycatch. Atlantic stingrays must be handled with care, because the venomous spine can inflict extremely painful wounds. For safety, bathers should shuffle their feet when walking in areas where stingrays may be buried.

Bluntnose stingray - *Dasyatis say* (Lesueur, 1817)

KEY FEATURES: Distance from eye to tip of snout shorter than distance between spiracles; front outline of disk wings weakly convex near tip of snout. COLOR: Brown on back, white below; fin-folds black. SIZE: Maximum 1 m (3.3 ft) DW. RANGE: New Jersey to Florida Keys and the northern and western Gulf of Mexico. Apparently absent from Central America. Another population occurs from Venezuela to Argentina.

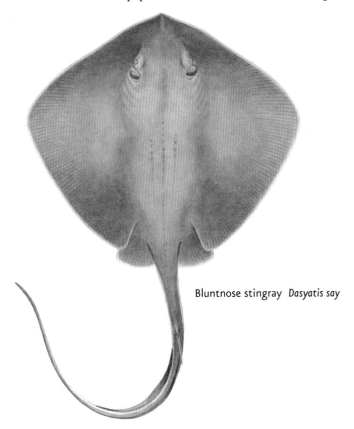

Bluntnose stingray *Dasyatis say*

HABITAT AND HABITS: Bluntnose stingrays make seasonal migrations into warm-temperate latitudes in summer and return to warmer climes for the winter. They lie partially covered with sediment on sandy or muddy bottoms. Bluntnose stingrays usually occur at depths less than 10 m (33 ft) but may range deeper in winter. OCCURRENCE IN THE CHESAPEAKE BAY: Seasonal visitors to the Chesapeake Bay during the summer and fall, bluntnose stingrays are the most common stingray in the bay. During late September to early October, they are occasionally abundant in the lower bay.

Although bluntnose stingrays prefer high-salinity waters, they are also occasionally found in salinities as low as 9‰.

REPRODUCTION: Bluntnose stingrays are viviparous, and the developing embryos are nourished by uterine "milk" secretions. Bluntnose stingray males mature at 30–40 cm (12–16 in) DW, females at 50–60 cm (20–24 in) DW. Gestation has been estimated at 9–10 months, and 2–4 pups are born in the late spring and summer. In Virginia, the pupping and nursery areas are in the inlets and lagoons of the Eastern Shore. Mating apparently takes place soon after pupping. Embryonic development may slow or stop during the winter.

FOOD HABITS: Bluntnose stingrays are reported to flap their fins to uncover benthic food items such as small shrimps, crustaceans, worms, and mollusks. They crush the latter with their broad, pavement-like dentition.

IMPORTANCE: There are no directed fisheries in the Chesapeake Bay for bluntnose stingrays, but they are taken as unwanted bycatch. Bluntnose stingrays must be handled with great care because the venomous spine can inflict extremely painful wounds. Stingrays are known to cause injuries to bathers who step on them.

Spiny butterfly ray - *Gymnura altavela* (Linnaeus, 1758)

KEY FEATURES: Distinct single long tentacle at inner posterior corner of each spiracle; disk at least 1.5 times wider than long; 1 or more venomous spines at very base of tail; pelvic fin narrow. **COLOR:** Brown to tan above, and mottled with numerous light and dark spots; creamy white below; tail may be slightly paler than disk and have alternating cross-bars. **SIZE:** In the western Atlantic, the maximum adult size is 2 m (6.6 ft) DW. However, there are reports of spiny butterfly rays reaching 4 m (13 ft) DW in the eastern Atlantic. **RANGE:** Known from tropical, subtropical, and warm-temperate waters along both the eastern and western Atlantic coasts. This species is most abundant in the tropical latitudes but is not common anywhere. In the western Atlantic, this species is seasonally known from Massachusetts to northern Argentina, with sporadic occurrences in the Gulf of Mexico and Central America.

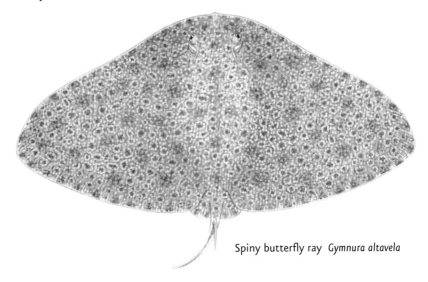

Spiny butterfly ray *Gymnura altavela*

HABITAT AND HABITS: Spiny butterfly rays inhabit sandy to muddy bottoms and occur from depths of less than 10 m (33 ft) to 100 m (325 ft). They migrate into warm-temperate latitudes in summer and return to warmer climes in winter. **OCCURRENCE IN THE CHESAPEAKE BAY:** These seasonal visitors to the lower Chesapeake Bay are present from May to November. During the summer, spiny butterfly rays occur frequently throughout the lower Chesapeake Bay and are especially common in the inlets of the Eastern Shore of Virginia.

REPRODUCTION: Spiny butterfly rays are viviparous, and the developing embryos are nourished by uterine "milk" secretions. Maturity occurs at about 1 m (3.3 ft) DW.

Two to seven pups are born in the summer after a six-to-nine month gestation period. In Virginia, spiny butterfly rays use the Eastern Shore inlets as pupping grounds.

FOOD HABITS: Spiny butterfly rays feed on all kinds of bottom-dwelling animals, but their diet consists primarily of small fishes and squids.

IMPORTANCE: Spiny butterfly rays are not known to be targeted by any fisheries, although occasional bycatch in trawl fisheries must occur. Care should be taken when handling this species because of its venomous spine. However, it is much less dangerous than the whiptail stingrays (Family Dasyatidae) because of its smaller spine, which is placed very close to the back of the disk and is therefore much less effective.

Smooth butterfly ray - *Gymnura micrura*
(Bloch and Schneider, 1801)

KEY FEATURES: Tentacle-like lobe absent on spiracle; disk at least 1.5 times wider than long; tail very short, less than 1/3 disk length, without spine. COLOR: Brown to tan above, with purplish or greenish iridescence, mottled with numerous light and dark spots, giving marbled appearance; creamy white below; well-defined dusky bands on tail. SIZE: The maximum adult size is 1.2 m (3.9 ft) DW; common to 90 cm

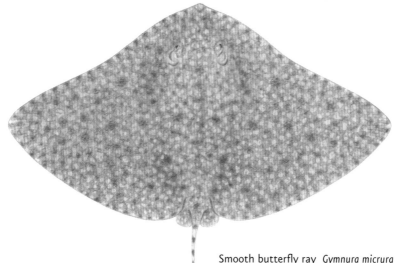

Smooth butterfly ray *Gymnura micrura*

(2.9 ft) DW. range: Tropical waters during the cooler months, subtropical to warm-temperate waters in warmer months. Known from the Chesapeake Bay to Brazil but apparently absent in the Antilles and common in the Gulf of Mexico.

habitat and habits: As with spiny butterfly rays, smooth butterfly rays inhabit sandy to muddy bottoms and occur from depths of less than 10 m (33 ft) to 100 m (325 ft). They migrate into warm-temperate latitudes in summer and return to warmer climes in winter. occurrence in the chesapeake bay: These seasonal visitors to the lower Chesapeake Bay are present from May to November, much like spiny butterfly rays. During the summer, smooth butterfly rays occur frequently throughout the lower Chesapeake Bay and are especially common in the inlets of the Eastern Shore of Virginia.

reproduction: Smooth butterfly rays are viviparous, and the developing embryos are nourished by uterine "milk" secretions. Maturity occurs at about 42 cm (16.5 in) DW in males and 50 cm (19.7 in) DW in females. Two to six pups are born in the summer after a six-to-nine month gestation period. In Virginia, smooth butterfly rays use the Eastern Shore inlets and lagoons as pupping and nursery grounds.

food habits: Smooth butterfly rays consume bivalve mollusks, crabs, shrimps and other crustaceans, and small fishes.

importance: Smooth butterfly rays are not targeted by any fisheries in the Chesapeake Bay region, although occasional bycatch in trawl fisheries must occur.

Bullnose ray - *Myliobatis freminvillei* Lesueur, 1824

key features: Distinct head and snout protruding well in front of disk margins. color: Grayish, reddish brown, or dark brown with diffuse whitish spots on back, white on bottom; teeth green. size: Maximum adult size 86 cm (2.8 ft) DW. range: Warm-temperate to subtropical waters from Delaware Bay (rarely from as far north as Cape Cod) to Florida and Gulf of Mexico. Separate population from Venezuela to Argentina, but records north of southern Brazil are suspect because of confusion with the closely related tropical species *Myliobatis goodei*.

habitat and habits: Bullnose rays are most common in shallow estuaries of 10 m (33 ft) or less in depth. Bullnose rays make seasonal migrations into warm-temperate latitudes in summer and return to warmer climes in winter. The population that summers in the Mid-Atlantic Bight migrates south of Cape Hatteras in the autumn.

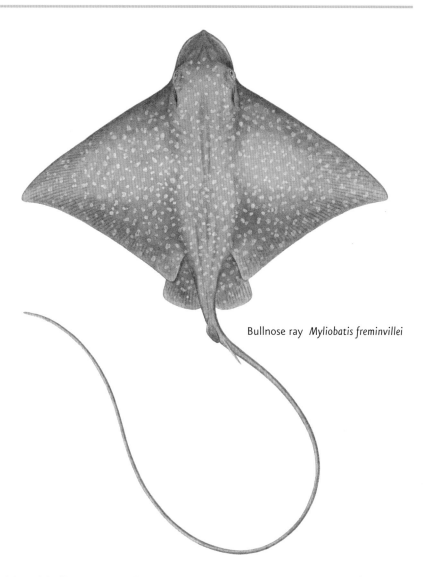

Bullnose ray *Myliobatis freminvillei*

Although bullnose rays are demersal, they have been known to leap out of the water.
OCCURRENCE IN THE CHESAPEAKE BAY: These seasonal visitors are occasional to common during the summer and fall in the lower Chesapeake Bay and rare in the upper Chesapeake Bay.

REPRODUCTION: Bullnose rays are viviparous, and the developing embryos are nourished by uterine "milk" secretions. Maturity occurs at 60–70 cm (24–28 in) DW, and 4–8 pups are born in the summer, usually in estuarine pupping and nursery grounds.

FOOD HABITS: Bullnose rays cruise over the bottom, rooting up crustaceans and mollusks with their protruding snout.

IMPORTANCE: There are no fisheries for bullnose rays in the Chesapeake Bay. Bullnose rays should be handled with care because of the venomous spine.

Cownose ray - *Rhinoptera bonasus* (Mitchill, 1815)

KEY FEATURES: Distinct head protruding in front of disk, with forehead divided into two lobes, and two fleshy lobes protruding below snout so that front of head appears to be four-lobed. COLOR: Various shades of brown on back; white below. SIZE: Maximum adult size 2.1 m (7 ft) DW, however, most western North Atlantic specimens are less than 1 m (3.3 ft) DW. RANGE: Warm-temperate to tropical waters from southern New England to Argentina. Apparently separate populations exist along the Atlantic Coast and the Gulf of Mexico. This species also occurs in the eastern Atlantic off northwest Africa.

HABITAT AND HABITS: Cownose rays migrate into the Mid-Atlantic Bight in late spring, where they remain for the summer. In autumn, they migrate south as far as the central Florida coast. Even though cownose rays feed on benthic prey, they often swim at the surface with their disk tips protruding rhythmically out of the water as they flap their wings. This habit has led to cownose rays often being mistaken for sharks. Cownose rays usually occur in small schools of a few to tens of individuals. However, they often form massive schools of hundreds of thousands or even millions, prior to and during their autumn migrations. OCCURRENCE IN THE CHESAPEAKE BAY: Cownose rays are abundant in the lower and middle Chesapeake Bay and into the brackish tributaries from late May to September, extending as far north in the bay as the Chester River. They may be found at any depth in the bay, from less than a meter (3.3 ft) deep to the deepest channels (approximately 30 m, or 98 ft).

REPRODUCTION: Cownose rays are viviparous, and the developing embryos are nourished by uterine "milk" secretions. Cownose rays pup in the Chesapeake Bay in June and July, about a month after their seasonal arrival. They may produce from one to (rarely) six pups and mate again soon after pupping. Gestation is about nine months, and embryonic development slows or stops in winter. The Chesapeake Bay serves as the principal pupping and nursery ground on the East Coast for this species. Maturity is reached at 61–71 cm (2–2.3 ft) DW.

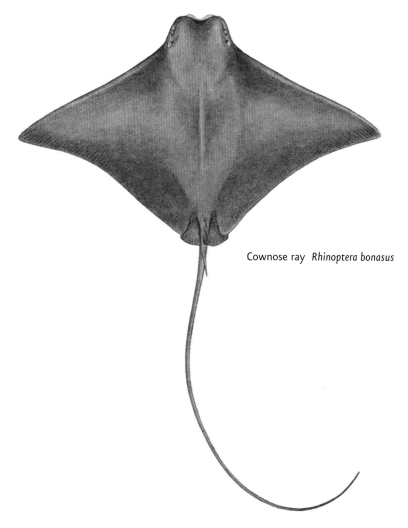

Cownose ray *Rhinoptera bonasus*

FOOD HABITS: Cownose rays feed on a variety of bivalve mollusks and other benthic prey. They often use their pectoral fins to excavate the bottom, in order to suck up sediment and winnow out burrowing clams or worms to eat. When feeding cownose rays move through an area, they may leave the bottom littered with craters measuring more than 1 m (3.3.ft) across and 0.5 m (1.8 ft) deep. This behavior can lead to habitat damage, particularly in beds of submerged aquatic vegetation.

IMPORTANCE: Cownose rays may inflict a significant toll on commercial shellfish beds. Consequently, even though there have not been any traditional fisheries for this species in the United States, Sea Grant scientists at the National Oceanic and

Atmospheric Administration have been striving to develop markets and fisheries for cownose rays as a means to reduce their abundance. Thus far this effort has met with limited success.

Barndoor skate - *Dipturus laevis* (Mitchill, 1818)

KEY FEATURES: Thorns absent along midline of disk. **COLOR:** Brown to tan to reddish brown on upper surface, with dark spots; white below, with gray patches and numerous dark pores. **SIZE:** Among the largest skates in the western North Atlantic, reaching a maximum adult size of 1.5 m (4.9 ft) TL. **RANGE:** Cold-temperate waters from Newfoundland and the Gulf of St. Lawrence to North Carolina.

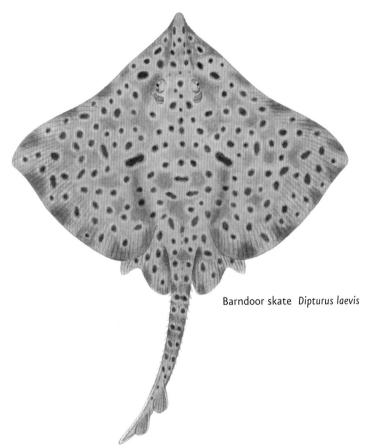

Barndoor skate *Dipturus laevis*

HABITAT AND HABITS: Barndoor skates occur from the subtidal zone in the northern parts of their range to the upper continental slope (200–700 m, or 650–2275 ft) in the south. They may make onshore migrations in the colder months in the Mid-Atlantic Bight and return offshore to the shelf edge in spring. Barndoor skates seem to have no strong preference for bottom type and have been found over rock, gravel, sand, and mud substrates as well as on sea scallop beds. OCCURRENCE IN THE CHESAPEAKE BAY: Formerly a common winter visitor in the lower Chesapeake Bay, barndoor skates have rarely been recorded in recent years.

REPRODUCTION: Barndoor skates mature at six to eight years of age. Females deposit 40–50 horny egg capsules on the bottom, 2 at a time every few days during the winter. The capsules hatch the following summer.

FOOD HABITS: Small barndoor skates consume mostly benthic invertebrates such as worms and crustaceans until they reach a size of about 70 cm (2.3 ft) TL, after which the diet and dentition of males and females begin to diverge. The diet of larger males is composed of about 20% crustaceans and mollusks and 80% fishes, including sea herring and a diversity of demersal fishes. Females split their diet about equally between fishes and crustaceans.

IMPORTANCE: Barndoor skates are not the target of either commercial or recreational fisheries in the Chesapeake Bay region.

Little skate - *Leucoraja erinacea* (Mitchill, 1825)
Winter skate - *Leucoraja ocellata* (Mitchill, 1815)

KEY FEATURES: In all size classes, little skates have fewer rows of teeth in the upper jaw than winter skates. For instance, in specimens greater than 35 cm (13.7 in) TL, little skates have ≤53 tooth rows in the upper jaw, whereas winter skates have ≥63.
COLOR: Both of these skates are grayish brown above, with numerous scattered dark spots, but winter skates often have paired "eye" spots on the pectoral fins as well.
SIZE: Little skates may reach 54 cm (1.8 ft) TL, but 41–51 cm (1.4–1.7 ft) is more common. Winter skate males may reach 110 cm (3.6 ft) TL, but the largest recorded female is 81 cm (2.7 ft) TL. RANGE: Both skates are found in cold-temperate and coastal waters. Little skates range from the southern Gulf of St. Lawrence to North Carolina, whereas winter skates occur from the Grand Banks of Newfoundland to North Carolina.

HABITAT AND HABITS: Both of these skates are benthic and prefer sand or gravel bottoms and water temperatures between 2° and 15°C (34–59°F). They move inshore and offshore seasonally to at least as deep as 90–120 m (293–390 ft) in order to remain within an acceptable thermal range. OCCURRENCE IN THE CHESAPEAKE BAY: Both skates are occasional winter or early spring visitors in the lower bay.

REPRODUCTION: Little skates reach maturity at 32–43 cm (1–1.4 ft) TL at an age of 3–4 years. Winter skates mature at 73–76 cm (2.4–2.5 ft) TL at an age of 11–12 years. Both species appear to reproduce at any time throughout the year, with individual little skates producing up to 35 horny eggs annually and winter skates depositing up to 40 eggs. Whereas little skates lay their eggs on sandy or even muddy bottoms, winter skates appear to prefer rocky bottoms.

FOOD HABITS: Little skates consume a wide variety of bottom prey, including small crabs, shrimps, and other crustaceans, worms, sea squirts, and clams. Squids and small fishes are also occasionally eaten. Winter skates eat crabs, fishes, and squids but also worms and various mollusks.

IMPORTANCE: There is no commercial or recreational fisheries interest for either of these skates in the Chesapeake Bay region. However, there are directed commercial fisheries for both species off New England.

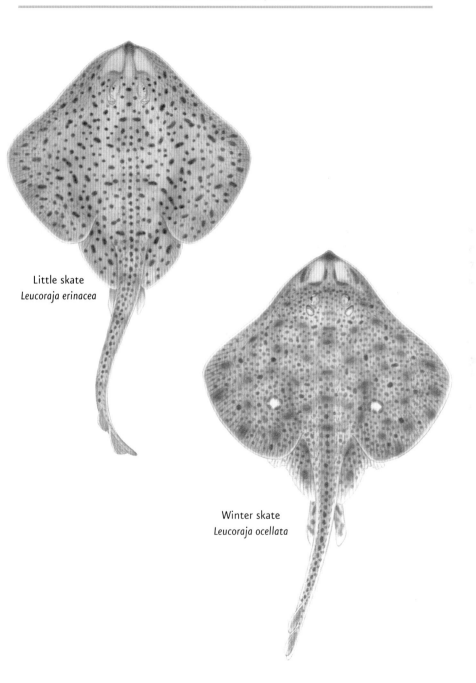

Little skate
Leucoraja erinacea

Winter skate
Leucoraja ocellata

Clearnose skate - *Raja eglanteria* Bosc, 1800

KEY FEATURES: Snout acutely pointed; area on either side of snout semitransparent; midrow thorns present from behind eyes to first dorsal fin near tail tip. **COLOR:** Pale brown dorsally, with distinctive dusky brown elongate spots and bars, white belly. **SIZE:** Maximum adult size 95 cm (3.1 ft) TL but typically much smaller. **RANGE:** Coastal warm-temperate waters from Massachusetts to Texas.

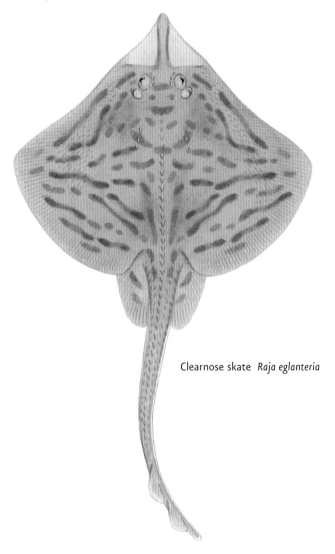

Clearnose skate   *Raja eglanteria*

HABITAT AND HABITS: Clearnose skates inhabit soft bottoms from the shoreline to depths as great as 300 m (975 ft). However, they are most common at depths less than 100 m (325 ft). In the Mid-Atlantic Bight, clearnose skates make migrations inshore and north in the spring and offshore and south to the outer continental shelf in the winter. They occur over a wide temperature range (10–30°C, or 50–86°F), but in the northern part of their range, they prefer water temperatures of 10–21°C (50–70°F). OCCURRENCE IN THE CHESAPEAKE BAY: Clearnose skates are common in the lower Chesapeake Bay from mid-spring to mid-autumn but may move into deeper bay waters or into nearshore coastal waters in mid-summer when water temperatures are high. They are absent or rare in winter.

REPRODUCTION: Clearnose skates are egg-layers and deposit up to 60 horny egg capsules per year in pairs on muddy to sandy bottoms. Incubation takes about three months. Males mature at 56 cm (1.8 ft) TL, females at 60 cm (2 ft) TL. Mating takes place inshore, and females may store sperm for at least three months.

FOOD HABITS: Clearnose skates feed mainly at night on crabs, shrimps, small clams, worms, small squids, and a variety of bottom-dwelling fishes.

IMPORTANCE: Considered a nuisance by both commercial and recreational fishermen. Caught incidentally with bottom trawls, with pound nets, and by rod and reel. There are no directed fisheries for this species, although it is frequently taken as discarded bycatch in trawl fisheries. The population appears to be healthy.

Shortnose sturgeon - *Acipenser brevirostrum*
Lesueur, 1818

KEY FEATURES: Large sucker-like mouth, mouth width 63–81% of distance between eyes; 5 rows of prominent bony scutes coursing length of body from behind head to base of tail; small individual scutes absent between anal fin and lateral scute row; single row of scutes before anal fin. **COLOR:** Dark brown to greenish above, white below; dusky pigment in anal fin. **SIZE:** These are the smallest of the sturgeons, attaining a maximum size of 1.4 m (4.6 ft); however, most are smaller. **RANGE:** Tidal freshwaters and low-salinity estuaries from St. John River, New Brunswick, to St. Johns River, Florida.

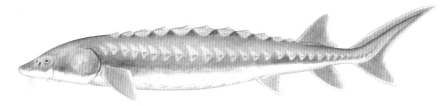

Shortnose sturgeon *Acipenser brevirostrum*

HABITAT AND HABITS: Shortnose sturgeon are usually resident in specific river systems and their downstream estuaries, rarely venturing out to the coast as do Atlantic sturgeon. Freshwater populations of shortnose sturgeon are known, but most are anadromous, moving from tidal freshwater into brackish water and back. **OCCURRENCE IN THE CHESAPEAKE BAY:** Historical records show that spawning populations of shortnose sturgeon were present at least until the late nineteenth century in the Potomac and Susquehanna Rivers. However, as in many other rivers, overfishing, pollution, and dam construction extirpated these populations by the early twentieth century. More recently, the U.S. Fish and Wildlife Service has documented occasional catches of shortnose sturgeon in the upper Chesapeake Bay, with individual records from the Potomac and Rappahannock Rivers. These recent records in all probability represent individuals that entered the upper Chesapeake Bay via the Chesapeake and Delaware Canal from the Delaware River, where the species has recovered and is common.

REPRODUCTION: Shortnose sturgeons ascend rivers to spawn in swift water, over rocky bottom above the fall line, from February to April. Males mature in three to five years, and females at six or seven years. Females may deposit from 27,000 to 208,000 eggs per season. Shortnose sturgeon can live to 30–40 years of age. The oldest known individual was 67 years of age.

FOOD HABITS: Shortnose sturgeons reportedly feed mostly at night over soft substrates. Juveniles feed primarily on benthic crustaceans and insects, whereas adults prey on benthic crustaceans, insects, and mollusks.

IMPORTANCE: Shortnose sturgeons are classified as an endangered species by the U.S. Fish and Wildlife Service and cannot be legally caught for commercial or recreational purposes. They are protected throughout their range. Protection and improvement in water quality have led to recovery of this species in some river systems. In addition, ongoing dam removal from some rivers where shortnose sturgeons survive will open critical spawning habitat. The U.S. Fish and Wildlife Service has reared shortnose sturgeon under hatchery conditions for several years, thus opening the door to reintroduction into rivers where the species has been extirpated. Reintroduction with hatchery fish may be the only way to restore spawning shortnose sturgeon populations to the Chesapeake Bay watershed.

Atlantic sturgeon - *Acipenser oxyrinchus* Mitchill, 1815

KEY FEATURES: Atlantic sturgeons may be distinguished from shortnose sturgeons by the small mouth (43–66% of the interorbital width, or distance between the eyes), 2–6 bony scutes between the anal-fin base and the lateral-scute row, a double row of dorsal scutes posterior to the dorsal fin, and a double row of scutes anterior to the anal fin. COLOR: Dark brown to greenish, sometimes with pinkish tinge above, white below. SIZE: Maximum adult size 4.3 m (14 ft) TL and more than 363 kg (800 lb). RANGE: Labrador, Canada, to St. Marys River, Florida.

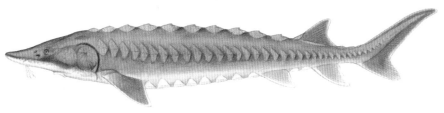

Atlantic sturgeon *Acipenser oxyrinchus*

HABITAT AND HABITS: Juvenile Atlantic sturgeons remain in the tidal freshwaters in which they were spawned for a year or longer before descending into brackish estuarine nurseries and then into the ocean. Their coastal sojourns, usually in winter and early spring, may span hundreds of kilometers but usually are shorter and restricted to depths less than 40 m (130 ft). Foraging Atlantic sturgeons often occur in estuaries and river mouths far from their natal rivers. OCCURRENCE IN THE CHESAPEAKE BAY:

Atlantic sturgeons once spawned in all the major tributaries of the Chesapeake Bay. Overfishing, pollution, and dam construction decimated the population in the late nineteenth and early twentieth centuries. Presently a recovering spawning population survives in the James River and to a lesser extent in the York River. Apparently, spawning populations in all the other tributaries have been extirpated. Regardless, juvenile and occasional adult Atlantic sturgeons occur regularly throughout the bay. Many of these are from other drainage systems, and tagged Atlantic sturgeons from the Hudson River and South Carolina have been taken in bay waters.

REPRODUCTION: Atlantic sturgeons enter the Chesapeake Bay in the early spring from their coastal wintering grounds and stage in the lower James River off Jamestown Island. As spring progresses, they move farther upstream and spawn somewhere below the fall line in May or early June. One spawning site is located near the junction of Turkey Island Oxbow and the main channel of the James River below Hopewell. Females produce 800,000 to 3.76 million eggs but usually do not spawn every year. The eggs are demersal and sticky and require clean hard substrate with some current for successful development. Males may remain in the spawning area into the autumn. Females usually move downstream soon after spawning, but some return for a second spawning event in the autumn. Females mature at 7–12 years, and males mature a bit earlier. Atlantic sturgeon can live for 60 years.

FOOD HABITS: Atlantic sturgeons feed over soft substrates, sucking benthic invertebrates such as mollusks, crustaceans, and aquatic insects from the sediment. Juvenile Atlantic sturgeons spend the day in deeper water at the edge of channels and move onto shallow flats to feed at night.

IMPORTANCE: Tragically, Atlantic sturgeons have been so reduced by overfishing, pollution, and dam construction that in 1998 the Atlantic States Marine Fisheries Commission implemented a coastwide moratorium on the harvest of Atlantic sturgeon that will remain in effect until there are at least 20 protected year-classes in each spawning stock (anticipated to take up to 40 or more years). In 2012, the National Marine Fisheries Service listed the Atlantic sturgeon population in the Chesapeake Bay as Endangered under the Endangered Species Act.

Longnose gar - *Lepisosteus osseus* (Linnaeus, 1758)

KEY FEATURES: Snout extremely long and beak-like; base of tail fin asymmetrical; body covered with bony tile-like scales. COLOR: Dark brown to olive on back, grading to tan on sides, and white on belly. Fins often with dark spots. Small juveniles with dark stripe along each side of body. SIZE: Maximum adult size 1.8 m (6 ft) TL; females generally larger than males. RANGE: Fresh and brackish waters along the Atlantic coast from Quebec to northern Mexico. Also found in inland waters from the Great Lakes to the Rio Grande.

Longnose gar *Lepisosteus osseus*

HABITAT AND HABITS: Longnose gars occur primarily in freshwater in rivers, streams, lakes, and swamps but also occur with regularity in brackish water with salinities as high as 10‰. Apparently, longnose gars can survive in full seawater (32‰) for at least limited periods of time, as they are occasionally captured in the ocean by beach seiners in North Carolina. Longnose gars often hover close to the surface and can gulp air when dissolved oxygen levels in the water are low in summer. The swim bladder in these primitive fish is attached directly to the gut and can absorb oxygen from air. Larval longnose gars have an adhesive disk on the tip of the snout which they use to attach to floating vegetation and other objects. Juvenile longnose gars grow quickly (>3 mm, or >0.1 in, per day) and may reach half their adult size or more in their first year. OCCURRENCE IN THE CHESAPEAKE BAY: Longnose gars are the principal piscivorous fish resident in tidal freshwaters throughout the Chesapeake Bay and its tributaries. They are abundant in spring, summer, and autumn but disappear in winter, when their behavior and whereabouts are virtually unknown.

REPRODUCTION: Longnose gars spawn in shallow freshwater in May and June. The same spawning sites appear to be used in successive years. Several males may escort a single female during spawning. Fecundity is 6,200–77,150 eggs. The adhesive eggs are dark green and supposedly toxic to humans. Males mature at 60 cm (23.6 in) TL and 2–3 years of age, whereas females mature at 80 cm (31.5 in) TL and 5–6 years of age. Males may reach a maximum age of 17 and females, 22.

FOOD HABITS: Longnose gars consume a wide variety of freshwater, estuarine, and marine fishes. In Chesapeake Bay tributaries, juvenile menhaden and white perch are regular prey items. Longnose gars are lie-in-wait predators, dashing out to capture passing prey with a swipe of their gator-like jaws.

IMPORTANCE: There are no commercial fisheries for longnose gars in the Chesapeake Bay, although a small number of recreational anglers target them for sport.

Bowfin (grinnel) - *Amia calva* Linnaeus, 1766

KEY FEATURES: Mouth large; anterior nares (nostrils) tubular, overhanging mouth; dorsal fin low and long-based, extending almost to tail; tail fin rounded and asymmetrical at base. **COLOR:** Dark brown to olive on back and sides, forming reticulated pattern; belly light tan to yellow or cream; two or three dusky stripes radiating from eye; ocellus (eye spot) at upper tail base, ocellus dark, ringed with red, orange, or yellow in adult males, typically absent in adult females. **SIZE:** Maximum adult size 1 m (3.3 ft) TL, commonly 40–82 cm (16–32 in) TL. **RANGE:** Freshwaters, mostly in the coastal plain, ranging from Connecticut to Texas and up into the Mississippi Valley to Michigan.

Bowfin *Amia calva*

HABITAT AND HABITS: Bowfin are found in blackwater swamps, lakes, and sluggish river backwaters and prefer shallow weedy habitats with stumps or other cover. They will occasionally be found in saline waters of 5‰ or greater. Bowfin have a lung-like swim bladder that allows them to breathe air and to survive under low-oxygen conditions, and they are capable of aestivation in mud under low-water conditions. **OCCURRENCE IN THE CHESAPEAKE BAY:** Bowfin occur in tidal freshwater tributaries but are uncommon in most locales, particularly in Maryland, where they are known from fewer than a dozen localities. They are common in Back Bay and the Dismal Swamp.

REPRODUCTION: Bowfin mature at 38–68 cm (15–27 in) TL, with males maturing at smaller sizes than females. Maximum age in nature is not well known, but captive specimens have survived for 30 years. Males build a saucer-shaped nest on the bottom near cover, and spawning occurs at night from April to June. Nests may contain 2,000–5,000 sticky demersal eggs. Individuals may spawn several times with different mates during the season, and females may produce as many as 64,000 eggs. Young bowfin possess attachment organs on their snouts that enable them to adhere to the nest or other objects. Males defend their nest and continue to defend their brood even after they leave the nest, until the fingerlings reach about 10 cm (4 in) TL.

FOOD HABITS: Bowfin primarily eat other fishes but will consume practically anything they can catch, including worms, crayfish, insects, mollusks, and frogs. Small juveniles eat microcrustaceans and insects.

IMPORTANCE: Bowfin are strong and tenacious fighters when taken on hook and line and were even privately stocked in some inland waters a few decades ago. Today, few anglers target bowfin, for despite their sporting qualities, their flesh is of very poor eating quality.

Ladyfish - *Elops saurus* Linnaeus, 1766

KEY FEATURES: Scales very small, 100–120 along lateral line; dorsal fin without elongate terminal ray. COLOR: Bright silvery on sides and belly, more bluish on back. SIZE: Maximum adult size 1 m (3.3 ft) TL. RANGE: Subtropical and tropical waters from Massachusetts (summer, rare) to Florida, Gulf of Mexico, and Caribbean to Brazil.

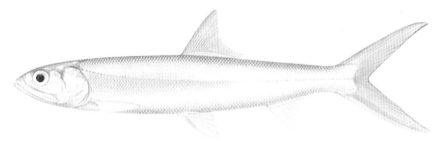

Ladyfish *Elops saurus*

HABITAT AND HABITS: Ladyfish occur in large schools in the warmer parts of their range and frequent inshore oceanic waters and estuaries, sometimes ascending into freshwater. Juveniles are common in mangrove creeks and lagoons. OCCURRENCE IN THE CHESAPEAKE BAY: Ladyfish are occasional summer visitors, usually in the lower Chesapeake Bay, but may be expected almost anywhere because of their wide salinity tolerance. They have been collected as far north in the Chesapeake Bay as the Nanticoke River.

REPRODUCTION: Ladyfish spawn in southern offshore waters in the late summer and autumn. The larvae, called leptocephali, are laterally flattened and ribbon-like and undergo complex metamorphic changes into the juvenile stage. Even though spawning is not known to occur north of Cape Hatteras, metamorphic larvae have been taken in May through July in low-salinity waters of the lower James River.

FOOD HABITS: Ladyfish are mostly piscivorous but also consume crustaceans.

IMPORTANCE: Ladyfish are not valued as food, and there is no commercial fishery in the Chesapeake Bay region.

Tarpon - *Megalops atlanticus* Valenciennes, 1847

KEY FEATURES: Dorsal fin with long trailing filament; 40–48 large scales along lateral line. COLOR: Uniformly silvery along sides and belly; silvery blue to green on back; pale pectoral and pelvic fins. SIZE: Maximum adult size 2.5 m (8.2 ft) TL. RANGE: Known from subtropical and tropical waters on both sides of the Atlantic Ocean. In the western Atlantic, tarpon range from Nova Scotia (rare north of Virginia) to Florida, throughout the Gulf of Mexico and Caribbean, and south to Argentina.

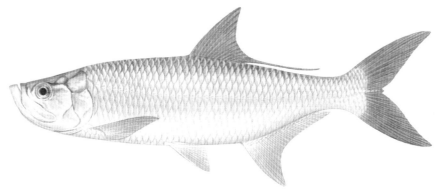

Tarpon *Megalops atlanticus*

HABITAT AND HABITS: Tarpon are marine fishes typically encountered no more than a few miles offshore. They have a wide salinity tolerance and often penetrate into freshwaters. Young tarpon are often found in estuaries and mangrove swamps. Tarpon typically frequent warm coastal waters, coral reefs, grass flats, and mangrove estuaries. Larger tarpon may migrate into warm-temperate waters in the summer. The swim bladder of tarpon is highly vascularized and lung-like; this allows them to supplement gill respiration by air-breathing. Breathing is accomplished by periodically gulping air at the surface while exhibiting a rolling behavior not unlike that of porpoises. This ability to gulp air and absorb oxygen in their lung-like swim bladder enables tarpon to withstand low-oxygen conditions sometimes found in their mangrove nurseries. OCCURRENCE IN THE CHESAPEAKE BAY: Tarpon visit the Chesapeake Bay in small numbers every summer and have been recorded as far north in the bay as the Choptank River. They are more common in the inlets and lagoons of the Eastern Shore of Virginia as well as in the tidal passes between the barrier islands.

REPRODUCTION: Tarpon spawn in southern offshore waters in the spring and summer. A typical female of 2 m (6.6 ft) TL may lay as many as 12 million pelagic eggs. The larvae, called leptocephali, are laterally flattened and ribbon-like and undergo complex metamorphic changes into the juvenile stage. Tarpon mature at about 1.2 m (4 ft) TL. Males may live for more than 30 years and females more than 50 years. Tarpon in captivity have reached 63 years of age.

FOOD HABITS: Juvenile tarpon feed on zooplankton and insects and include small fishes in their diet as they grow. Larger tarpon are voracious predators, engulfing whole fish, shrimps, and crabs.

IMPORTANCE: Tarpon are legendary gamefish. Their acrobatics when hooked have earned them the name of "Silver King" in the angling community. Care must be taken when landing large tarpon, even for release, as numerous instances are documented of tarpon leaping into boats and wreaking havoc on gear and people alike. Boated fish should be properly respirated prior to release in order to prevent death. No commercial fishery exists for tarpon, and the flesh is considered coarse and unpalatable to American taste.

American eel - *Anguilla rostrata* (Lesueur, 1817)

KEY FEATURES: Dorsal fin originates in front of midbody and runs continuously around tail, contiguous with anal fin; lower jaw projects slightly beyond snout; small scales embedded in skin. COLOR: Olive to dark brown above, whitish below; mature adults have bronze iridescence during autumn downstream spawning migration. SIZE: Females grow larger than males. The maximum adult size for females is 1.5 m (5 ft) TL, whereas males can reach 61 cm (2 ft) TL. RANGE: Present in brackish waters and their freshwater tributaries along the Atlantic coast of North America, throughout the Gulf of Mexico and the Caribbean, and along the east coast of Central America to Venezuela. They also occur inland in the St. Lawrence and the Great Lakes.

American eel *Anguilla rostrata*

HABITAT AND HABITS: American eels are catadromous; they spawn in the sea but enter estuaries and freshwater to feed and grow. Baby transparent "glass eels" darken into "elvers" as they arrive in the spring, and thousands may choke small coastal streams, even ascending small dams. The fresh and brackish water "yellow eel" phase lasts more than 5 years and may be as long as 43 years. Males predominate in brackish areas, whereas females ascend into freshwater and may occupy streams, rivers, lakes, swamps, and virtually any other accessible habitats, far inland. OCCURRENCE IN THE CHESAPEAKE BAY: American eels are common to abundant in all estuarine and fresh water habitats where high dams do not block their passage. Doubtless their former distribution, before the construction of dams, extended into the mountainous regions of all the major drainages in the Chesapeake Bay.

REPRODUCTION: Males mature at 22–44 cm (0.8–1.4 ft) TL and typically at ages of 3–10 years. Females mature at 37–100 cm (1.3–3.3 ft) TL and ages of 4–18 years. American eels undergo a metamorphosis as they approach maturity and begin their autumn spawning migration downstream. They cease feeding, their eyes and pectoral fins enlarge, and their skin takes on a metallic brassy sheen. After entering the ocean, they swim quickly across the continental shelf into oceanic waters and head for the Sargasso Sea north of the Bahamas, where they spawn in February at great depths. Individual females may produce 400,000 to 2.5 million eggs. American eels presumably die after spawning. The pelagic eggs develop into ribbon-like leptocephali that may grow and drift in the Gulf Stream for up to a year before metamorphosing into glass eels, which swim to the coast and enter fresh and brackish waters to perpetuate their life cycle.

FOOD HABITS: American eels are nocturnally active and have a diverse diet that includes worms, crustaceans, aquatic insects, snails, clams, and a wide variety of fishes. They may also scavenge fresh dead food.

IMPORTANCE: American eels are sought commercially throughout their range and are taken in a variety of "racks," fyke nets, and pots. All phases, from glass eels to elvers, yellow eels, and maturing "bronze eels," are harvested. Younger stages may be fried whole or held in culture to grow and be harvested at larger sizes. Yellow and bronze eels may be consumed fresh or smoked. In recent years, American eels have shown signs of decline in the Chesapeake Bay and throughout their range. The reasons for this decline, whether over-harvesting, habitat destruction, or poor larval survival, are not clear at present. A recently discovered parasite in the American eel's swim bladder is suspected as being partly responsible for the decline.

Conger eel - *Conger oceanicus* (Mitchill, 1818)

KEY FEATURES: Snout projects slightly beyond lower jaw; dorsal fin originates just behind pectoral fin and runs continuously around tail, contiguous with anal fin; body lacks scales. **COLOR:** Dark gray above, whitish below; dark margin on continuous dorsal and anal fins. **SIZE:** Maximum adult size 2.2 m (7.2 ft) TL; weight can exceed 40 kg (88 lb). **RANGE:** Inhabit coastal waters in the western Atlantic from Georges Bank and Massachusetts to Florida and, occasionally, from Florida to Texas in the Gulf of Mexico. Also known from the northeastern Atlantic.

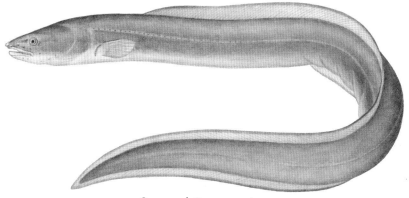

Conger eel *Conger oceanicus*

HABITAT AND HABITS: Conger eels are benthic cool-water species and occur from the shore out to 475 m (1545 ft). They may be resident on the continental shelf or may undertake seasonal migrations inshore and offshore. They are often associated with tilefish burrows at the shelf edge in the Mid-Atlantic Bight. **OCCURRENCE IN THE CHESAPEAKE BAY:** Conger eels are occasional visitors to the lower Chesapeake Bay in cooler months, extending as far up the bay as the mouth of the Potomac River on the Western Shore and the Nanticoke River on the Eastern Shore.

REPRODUCTION: Maturing conger eels depart from the continental shelf in summer and spawn in the Sargasso Sea north of the Bahamas in fall and winter. Adults apparently die after spawning. The oceanic leptocephali, transported by the Gulf Stream, undergo four basic morphological stages before metamorphosing at about 10 cm (4 in) TL and shrinking in length to become elvers. The baby eels become benthic on the continental shelf and may enter the mouths of estuaries. However, they are cryptic and have proven difficult to collect.

FOOD HABITS: Conger eels are nocturnal predators that are primarily fish eaters, but they will also prey on crustaceans and mollusks.

IMPORTANCE: There are no directed fisheries for conger eels in the Chesapeake Bay, but they are sometimes taken by recreational anglers while bottom fishing. Conger eels are also taken in the commercial trawl and pot fisheries as bycatch. The meat is palatable, particularly if smoked, and historically was marketed.

Blueback herring - *Alosa aestivalis* (Mitchill, 1814)

Alewife - *Alosa pseudoharengus* (Wilson, 1811)

KEY FEATURES: Blueback herring have smaller eyes, the diameter equal to or less than snout length; eye diameter of alewives is greater than snout length. **COLOR:** Blueback herring tend to be bluer on back than alewives, which have a greener back, but this can be variable, particularly after death. The peritoneum is black in blueback herring but pale with dusky spots in alewives. **SIZE:** Maximum adult size in both is 38 cm (1.3 ft) SL, but blueback herring are typically smaller than alewives. **RANGE:** Blueback herring occur from Nova Scotia to Florida, while alewives occur from Newfoundland to South Carolina. Where they overlap, blueback herring tend to dominate in the southern part of the range and alewives tend to dominate in the northern part. Landlocked populations of alewives occur in the Great Lakes (introduced) and

Blueback herring *Alosa aestivalis*

Alewife *Alosa pseudoharengus*

Finger Lakes and in several other smaller freshwater bodies. Some landlocked populations of blueback herring are known in the Southeast but are rare.

HABITAT AND HABITS: Both occur in schools, sometimes together. After hatching in freshwater, juveniles remain in freshwater nurseries in late spring and summer, moving downstream into higher salinities in the estuary in fall. Most young-of-the year migrate into the coastal ocean in winter, although some may remain in deeper portions of the bay until the following spring. Both species migrate into New England waters in summer and overwinter on the continental shelf as far south as Cape Hatteras. They follow this seasonal pattern for two to six years until they reach sexual maturity, when they return to their natal streams to spawn in spring. After spawning, surviving adults quickly return to sea, where they resume their coastal migration pattern. **OCCURRENCE IN THE CHESAPEAKE BAY:** Adult alewives appear in the lower bay in late winter and early spring and ascend freshwater tributaries to spawn in March and April. Blueback herring follow about three to four weeks later. Young-of-the-year persist in the estuary into late fall.

REPRODUCTION: Alewives mature in two to five years and ascend into freshwater to spawn in shallow areas with low flow or even in slightly brackish ponds. Blueback herring mature in three to six years and usually spawn in deeper areas with more flow than do alewives. Both species spawn in tidal and nontidal habitats and may spawn above the fall line if no barriers exist.

FOOD HABITS: Both are planktivorous. Juveniles feed mostly on microzooplankton and take larger prey as they grow. Adults feed on ctenophores, copepods, mysid shrimps, amphipods, and even small fishes.

IMPORTANCE: These two species are often taken together, and fishery statistics lump them together as "river herring." Historically, the fishery for river herring has been very important in the Chesapeake Bay. In modern times, most of the harvest has come from pound nets, but also from gill nets and haul seines. The landings in Maryland peaked in the 1930s, exceeding 3,600 mt (8 million lb) annually. The Maryland fishery has shown a steady decline, and in 2008, only 36 mt (80,000 lb) were landed. The fishery persisted longer in Virginia, with landings exceeding 13,600 mt (30 million lb) in the 1960s, but by 2008, landings had declined to 59 mt (130,000 lb). Degradation and destruction of spawning and nursery habitat, as well as dam construction, have been major causes of river herring declines. In addition, large harvests of river herring by foreign factory trawlers on the continental shelf during the late 1960s and early 1970s contributed to subsequent stock declines. Besides commercial fisheries, modest recreational fisheries have targeted river herring since colonial times. Fishers use large dip-nets in the smaller tributaries to capture river herring as they move upriver to spawn, usually at night. River herring are also taken by anglers using small gold spoons or spinners or jigging with small bare gold hooks. River herring also eagerly take small bright artificial wet flies.

Hickory shad - *Alosa mediocris* (Mitchill, 1814)

KEY FEATURES: Underslung lower jaw prominent, extending well beyond snout; upper jaw with deep median notch; in adults, 18–23 gill rakers on lower limb of first gill arch (fewer than in river herring and American shad). COLOR: Gray green above, silvery on sides, with dark shoulder spot, followed by series of smaller, more obscure spots. SIZE: Maximum adult size 60 cm (2 ft) SL but rarely greater than 45 cm (1.5 ft) SL. RANGE: Commonly from Cape Cod to northern Florida but known as far north as New Brunswick, Canada.

HABITAT AND HABITS: Hickory shad are anadromous and ascend freshwaters during the spring to spawn. Juveniles quickly return downstream to estuarine and coastal waters after metamorphosis, avoiding the freshwater nursery areas used by other shad species. Hickory shad remain in the ocean, close to the coast, often in the surf zone, for two to three years before returning to freshwater to spawn. Their north-south seasonal movements are limited compared to the other shads and herrings, and Chesapeake Bay hickory shad appear to spend their life in the inshore waters

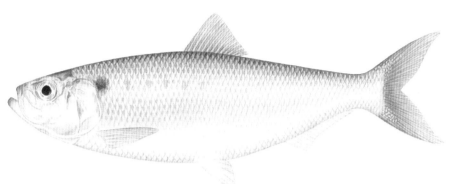

Hickory shad *Alosa mediocris*

of the Mid-Atlantic Bight. OCCURRENCE IN THE CHESAPEAKE BAY: Hickory shad ascend the larger tributaries in the Chesapeake Bay to spawn from April to early June and apparently move out to sea soon after spawning. Spawning grounds extend as far north in the Chesapeake Bay as the Susquehanna River. Juveniles spend little time in the Chesapeake Bay and descend tributaries to the lower Chesapeake Bay and ocean by June or July.

REPRODUCTION: Hickory shad in the Chesapeake Bay spawn mostly at or just below the fall line in the major tributaries. Substantial runs currently occur in the James, Rappahannock, Potomac, and Patuxent Rivers as well as on the Susquehanna flats.

FOOD HABITS: Hickory shad have a larger mouth and fewer gill rakers than other shads and are mostly piscivorous, feeding on a wide variety of fishes, including sand lances, anchovies, herring, and silversides. Their diet may also include squids and crustaceans.

IMPORTANCE: Hickory shad have rarely been the target of commercial fisheries in the Chesapeake Bay but have occurred as bycatch in other fisheries for shad and river herrings. In contrast, there is an intense, but brief, recreational fishery for hickory shad every spring at the fall line on the James, Rappahannock, and Potomac Rivers. Most anglers use spinning tackle and small gold spoons or brightly colored jigs to catch the fish, primarily for the sport, but also for the delicious roe. In recent years the fly fishing community has joined the action, using sink-tip lines and small chartreuse or fluorescent-orange flies.

American shad - *Alosa sapidissima* (Wilson, 1811)

KEY FEATURES: Eye diameter less than greatest cheek depth; in adults, 59–76 gill rakers on lower limb of first gill arch (many more than in the other local shad and river herring). COLOR: Gray green above, silvery on the sides, with dark shoulder spot sometimes followed by a few smaller, more obscure spots. SIZE: Maximum adult size 60 cm (2 ft) SL, rarely larger. RANGE: American shad occur from the St. Lawrence River, Canada, to northern Florida. They have been successfully introduced to the Pacific coast of the United States.

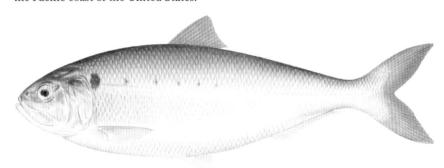

American shad *Alosa sapidissima*

HABITAT AND HABITS: American shad are an anadromous species that spends most of its adult life at sea in large schools. As juveniles, American shad remain in freshwater nurseries through the summer and into the early fall. They migrate to the ocean in late fall and overwinter on the continental shelf. In summer, American shad migrate from the Mid-Atlantic Bight up the coast into the Gulf of Maine and Bay of Fundy, where they may form large feeding schools. They return south to overwinter in offshore waters from Maryland to Cape Hatteras. Both juveniles and adults follow the same approximate migration patterns, except that mature fish return each spring to their natal rivers to spawn. OCCURRENCE IN THE CHESAPEAKE BAY: Adult shad may enter the Chesapeake Bay as early as January; however, spawning does not begin until March and may extend into early June, depending on the tributary. If they survive spawning, adults return to the sea within a few days to a few weeks afterward.

REPRODUCTION: Spawning occurs in the evening or at night, over shallow flats in both tidal and nontidal freshwater, and may extend far above the fall line if no barriers exist. Female fecundity is high (100,000–600,000 eggs). Survival after spawning and repeat spawning in subsequent years was thought to be rare in Chesapeake Bay shad, but this has recently been found to be untrue. The percent of repeat spawners varies by year and tributary and may exceed 25% in some instances.

FOOD HABITS: American shad are planktivorous, with small shad taking smaller prey, such as copepods, and larger shad feeding on larger prey, such as shrimps, jellyfishes, small fishes, and fish eggs.

IMPORTANCE: The shad fishery was the most important fishery in the Chesapeake Bay during the later 1800s, with landings reaching 7,700 mt (17 million lb) in 1900. Pollution, habitat destruction, and dams decimated the shad spawning grounds and nursery areas, leading to precipitous declines in shad stocks and fishery landings. Shad fisheries in the Chesapeake Bay were closed in Maryland in 1980 and in Virginia in 1994. Water quality in shad spawning and nursery areas has been improving over the last few decades due to federal clean water legislation. In addition, some dams on the James and Rappahannock Rivers have been removed, allowing passage of anadromous fishes into former spawning and nursery areas blocked for decades or longer. It remains to be seen whether all these restoration efforts can bring shad populations back to their former levels of abundance. American shad are eagerly sought in freshwater recreational fisheries throughout their range because of their size (regularly up to 2.7 kg, or 6 lb) and sporting qualities. Their habit of leaping while hooked has earned them the name "poor man's salmon."

Atlantic menhaden - *Brevoortia tyrannus*
(Latrobe, 1802)

KEY FEATURES: Head large; gill rakers long and slender; scales on midline in front of dorsal fin enlarged. **COLOR:** Back greenish to bluish, brassy on sides, large dark shoulder spot with several series of small spots on flanks. **SIZE:** Maximum adult size 38 cm (1.3 ft) SL. **RANGE:** Coastal waters from Nova Scotia to Florida.

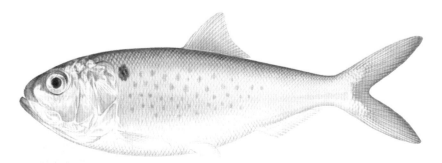

Atlantic menhaden *Brevoortia tyrannus*

HABITAT AND HABITS: Atlantic menhaden overwinter south of Cape Hatteras and migrate into the mid-Atlantic in the spring. They spend the summer in estuarine and coastal waters as far as the Gulf of Maine, with the largest, oldest fish migrating the farthest north. In autumn, menhaden return south, with most of the population off the Carolinas by December. Some young-of-the-year may overwinter in coastal waters of the Mid-Atlantic Bight. **OCCURRENCE IN THE CHESAPEAKE BAY:** Young menhaden usually enter the Chesapeake Bay in May and spread to virtually all habitats, from the mainstem to tidal freshwater. Larvae are wafted up into the tributaries from the Chesapeake Bay mouth by the salt wedge. The metamorphosed juveniles frequent tidal creeks, and larger juveniles occur in large schools in open water of the bay. Adults may enter the lower bay in the spring but continue north for the summer. Most menhaden leave the Chesapeake Bay and migrate south in the fall, but some juveniles may remain in deep channels of the bay, especially in mild winters.

REPRODUCTION: Most Atlantic menhaden mature at age three, and spawning occurs over the continental shelf near shore during most months of the year, with a strong peak in fall and winter and a secondary peak in spring. The pelagic eggs and larvae drift with the prevailing currents until the larvae recruit into the estuarine nurseries.

FOOD HABITS: Atlantic menhaden are filter feeders on both phytoplankton and zooplankton.

IMPORTANCE: Atlantic menhaden are used for bait, fertilizer, fish meal, and fish oil. More pounds of menhaden are landed each year in the United States than any other fish. Over the past three decades, the landings of menhaden in Virginia have averaged 230,000 mt per year (507 million lb) with a high of 320,000 mt (705 million lb) in 1990. Over this same period, the catch of menhaden in Maryland (where purse seines are prohibited) averaged 2600 mt per year (5.7 million lb) with a high of 7000 mt (15.8 million lb) in 2005. Currently the Atlantic States Marine Fisheries Commission sets an annual quota on the catch of menhaden. Atlantic menhaden are an extremely important forage species for a wide variety of fishes, including many species of recreational and commercial importance.

Atlantic herring - *Clupea harengus* Linnaeus, 1758

KEY FEATURES: Scales on belly not sharp and keeled; belly rounded; body depth 20–26% of SL, much narrower than all other local herrings, save the round herring, which has no scales on belly; 9 pelvic-fin rays (rarely 8 or 10). **COLOR:** Blue or greenish above and silvery on sides; no dark spot on shoulder. **SIZE:** Maximum adult size 40 cm (1.3 ft) SL; typically 20–25 cm (8–10 in). **RANGE:** Known from both coasts of the North Atlantic. In the western Atlantic, the range is from southern Greenland to South Carolina.

Atlantic herring *Clupea harengus*

HABITAT AND HABITS: A cold-temperate pelagic schooling species, Atlantic herring include several populations with different spawning and wintering areas. Atlantic herring spend the summer at higher latitudes and migrate south in winter. Their occurrence in the middle Atlantic is confined to the colder months. These fish migrate to Georges Bank in spring, where they feed on the zooplankton bloom. Juveniles may use estuaries as nurseries. **OCCURRENCE IN THE CHESAPEAKE BAY:** Atlantic herring are found in the Chesapeake Bay during winter and early spring and extend as far north as the Susquehanna flats; however, they are more abundant in the lower bay.

REPRODUCTION: Atlantic herring spawn on offshore banks from Georges Bank north, with spawning peaks in spring and fall. These fish often form large spawning schools and deposit their adhesive eggs on hard bottoms or occasionally on aquatic vegetation. Populations are subject to huge annual recruitment fluctuations, mitigated by the availability of plankton blooms to newly hatched larvae.

FOOD HABITS: Atlantic herring are filter feeders on plankton at all life stages. Small juvenile Atlantic herring feed on microzooplankton and phytoplankton. Larger herring feed on a wide variety of zooplankton, mainly larger copepods.

IMPORTANCE: Atlantic herring are of little importance to the fisheries of the Chesapeake Bay, as they are only occasional seasonal visitors. As with most species of this family (Clupeidae), Atlantic herring are important forage fishes in the ecosystems where they abound.

Gizzard shad - *Dorosoma cepedianum* (Lesueur, 1818)

Threadfin shad - *Dorosoma petenense* (Günther, 1867)

KEY FEATURES: Gizzard shad have 52–70 small scales along side of body, 29–35 anal-fin rays, and 17–19 sharp scales preceding pelvic fins, while threadfin shad have 41–48 large scales along side of body, 24–28 anal-fin rays, and 16–17 sharp scales preceding pelvic fins. **COLOR:** Gizzard shad have several series of dusky longitudinal stripes on flanks, not present in threadfin shad. Threadfin shad have a dark shoulder spot at all sizes, while in gizzard shad, a shoulder spot is present only in juveniles. The paired fins and tail are dusky to black in gizzard shad but yellowish in threadfin shad. **SIZE:** Gizzard shad may reach 50 cm (1.6 ft) SL, while the smaller threadfin shad only reach 22 cm (9 in) SL. **RANGE:** Gizzard shad are native to freshwaters of the Great Lakes, the Mississippi River and its tributaries, and fresh and brackish wa-

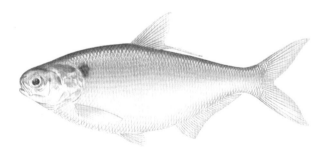

Gizzard shad *Dorosoma cepedianum*

Threadfin shad *Dorosoma petenense*

ters of drainages of the Gulf of Mexico and Atlantic coast as far north as New York. Threadfin shad are native in freshwaters from the lower Ohio Valley and in Mississippi drainages south to the Gulf of Mexico to Guatemala and Belize, but they readily enter brackish water. They have been widely introduced as forage fish, including into tributaries of the Chesapeake Bay.

HABITAT AND HABITS: Both these pelagic schooling shads occur in fresh to brackish waters (rarely marine) and occupy a variety of habitats, including large tidal rivers, streams, reservoirs, and lakes. Both are found over a diversity of substrate types, but gizzard shad may tolerate silty substrates better. In streams, threadfin shad prefer areas of faster current than do gizzard shad. OCCURRENCE IN THE CHESAPEAKE BAY: Gizzard shad are common to locally abundant in freshwater and slightly brackish habitats throughout the bay system, with some fish moving into deeper, more saline areas (ca. 20‰) in the late fall and winter. Threadfin shad were first introduced into reservoirs in Chesapeake Bay drainages in the 1950s and since have spread downstream into the major tributaries of the bay.

REPRODUCTION: Both mature at two to three years of age and spawn in shallow freshwater in spring and summer, with a peak from April to June. Both species have demersal adhesive eggs, which may stick to almost any submerged material, including aquatic vegetation, brush, rocks, and man-made structures.

FOOD HABITS: Both of these shads are filter-feeding planktivores, consuming both phytoplankton and zooplankton and sometimes bottom detritus.

IMPORTANCE: Neither has any value as a food or sport fish, but threadfin shad, in particular, have been widely stocked as forage fish in reservoirs and lakes. Large gizzard shad can be formidable opponents when foul-hooked on a fly rod, but their planktivorous diet has made suitable fly development a challenge that has yet to be met.

Round herring - *Etrumeus teres* (DeKay, 1842)

KEY FEATURES: The only herring species in the Chesapeake Bay region with pelvic fins located behind instead of beneath dorsal fin; pectoral and pelvic fins with prominent axillary process; belly is rounded and lacks scales. **COLOR:** Olive green above, with silvery sides and white belly. **SIZE:** Maximum adult size 33 cm (1.1 ft) TL; common from 18–25 cm (7–10 in). **RANGE:** Circumglobal in warm-temperate and subtropical nearshore regions. In the western Atlantic from Cape Cod (stray as far north as the Bay of Fundy) to Florida, in portions of the Gulf of Mexico and the Caribbean Sea, and southward to Venezuela and the Guianas.

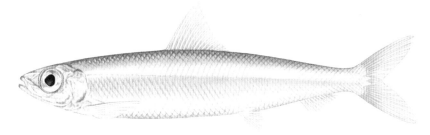

Round herring *Etrumeus teres*

HABITAT AND HABITS: Round herring occur over the continental shelf, migrating inshore in the late winter and spring and offshore in the summer and fall. They are confined to the deeper waters of the outer continental shelf in the southern part of their range. This species sometimes forms massive schools with other schooling fishes of similar size. Round herring concentrate close to the surface at night and descend into deeper waters during the day. **OCCURRENCE IN THE CHESAPEAKE BAY:** Round herring appear inshore off the Virginia coast in spring in large schools, where they may be consumed by juvenile humpback whales and various piscivorous fishes, including bluefish, summer flounder, and smooth dogfish. Round herring occur occasionally in spring in the lower Chesapeake Bay.

REPRODUCTION: Round herring have pelagic eggs and larvae, and they spawn over the middle and outer continental shelf in winter to spring.

FOOD HABITS: Round herring feed on zooplankton (mostly crustaceans) and fish larvae.

IMPORTANCE: Round herring are unimportant in Chesapeake Bay fisheries but are important forage species wherever they occur in abundance.

Atlantic thread herring - *Opisthonema oglinum*
(Lesueur, 1818)

KEY FEATURES: Gill rakers on first arch numbering fewer than 120; scales present on back in front of dorsal fin; single short dorsal fin with elongate trailing filament. COLOR: Bluish green on back and silvery on sides, with dark shoulder spot and several dusky horizontal stripes on upper flanks. SIZE: Maximum adult size 38 cm (1.3 ft) SL, commonly to 20 cm (8 in) SL. RANGE: Gulf of Maine (rarely) to the Gulf of Mexico, Bermuda, and the Caribbean south to southern Brazil.

Atlantic thread herring *Opisthonema oglinum*

HABITAT AND HABITS: Atlantic thread herring are a pelagic schooling species that frequents waters less than 50 m (163 ft) deep. Adult and juvenile Atlantic thread herring form schools near the surface, often with other herrings. Atlantic thread herring migrate north into Mid-Atlantic Bight nearshore waters in late spring and summer, returning to the South Atlantic Bight, the Gulf of Mexico, and tropical areas in the autumn. OCCURRENCE IN THE CHESAPEAKE BAY: Atlantic thread herring are common in the lower Chesapeake Bay in summer and migrate out of the bay to coastal waters south of Cape Hatteras in the fall.

REPRODUCTION: Atlantic thread herring have pelagic eggs and larvae and spawn in nearshore coastal waters in early summer.

FOOD HABITS: Atlantic thread herring are filter feeders that consume zooplankton, including copepods and other crustaceans, as well as small fishes.

IMPORTANCE: Atlantic thread herring are not harvested directly in the Chesapeake Bay but appear in pound-net catches, where they may be retained with a mixture of other small fishes for crab-pot bait. As with other small herrings, Atlantic thread herring are an important forage species wherever they occur in abundance.

Striped anchovy - *Anchoa hepsetus* (Linnaeus, 1758)
Bay anchovy - *Anchoa mitchilli* (Valenciennes, 1848)

KEY FEATURES: The anal fin of bay anchovies originates perpendicular to the front or center of the dorsal fin, whereas the origin of the anal fin in striped anchovies is further back, under the rear of the dorsal fin. In addition, bay anchovies have 23–31 anal-fin rays and 11–14 pectoral-fin rays, whereas striped anchovies have 18–24 anal-fin rays and 15–18 pectoral-fin rays. **COLOR:** Both have a silvery stripe along the side of the body, but this is broader and more prominent in striped anchovies. **SIZE:** Maximum adult size for bay anchovies is 10 cm (3.9 in) SL and for striped anchovies, 15 cm (5.9 in) SL. **RANGE:** Bay anchovies are found from the Gulf of Maine to the Gulf of Mexico, while striped anchovies occur from Massachusetts through the Caribbean to Uruguay.

HABITAT AND HABITS: Bay anchovies are primarily estuarine, occurring from tidal freshwater creeks to coastal marine waters, whereas striped anchovies are more restricted to higher estuarine and marine salinities. Both species form large schools. **OCCURRENCE IN THE CHESAPEAKE BAY:** Bay anchovies are the most abundant fish in the Chesapeake Bay and may be encountered in almost any habitat in the warmer months, from shallow marsh creeks to the bay mouth. In winter, bay anchovies move into the deeper waters of the tributaries and mainstem and into near coastal waters. Striped anchovies are much less common and are usually restricted to the higher salinities near the Chesapeake Bay mouth during the warmer months. In winter they move into coastal waters as deep as 70 m (228 ft).

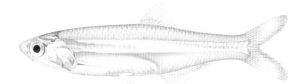

Striped anchovy *Anchoa hepsetus*

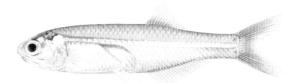

Bay anchovy *Anchoa mitchilli*

REPRODUCTION: Bay anchovies spawn over a wide range of salinities, usually >8–10‰ in the Chesapeake Bay from April to September, with a peak in July. Spawning is most intense in the evening, and individual females spawn every one to four days. Eggs and larvae are pelagic. Spawning of striped anchovies is restricted to the Chesapeake Bay mouth and coastal waters and also extends from April to September. Growth is rapid, with maturity occurring in three months to a year.

FOOD HABITS: Both of these anchovies consume zooplankton by filtering water with their very fine gill rakers.

IMPORTANCE: There are no fisheries for either of these very small species in the Chesapeake Bay, but both, particularly bay anchovies, are important forage species for a wide variety of piscivorous fishes, such as summer flounder, bluefish, striped bass, and weakfish. Anchovies are also important forage for a wide variety of fish-eating birds, including terns, gulls, herons, and egrets. Anchovies are an important component in the Chesapeake Bay ecosystem and are subject to large yearly population fluctuations mitigated by environmental conditions.

Quillback - *Carpiodes cyprinus* (Lesueur, 1817)

KEY FEATURES: Lateral line complete, 35–40 lateral line scales; dorsal-fin origin nearer to tip of snout than caudal-fin base; dorsal fin sickle-shaped (anterior rays long, posterior rays short) and long (with 24–33 fin rays), extending back over anal fin. **COLOR:** Olive above, brassy to silvery on sides and below; ventral fins gray or orangish with white leading edge. **SIZE:** Maximum adult size 66 cm (2.2 ft) TL. Com-

Quillback *Carpiodes cyprinus*

mon to 38 cm (1.3 ft) TL. RANGE: Quillbacks occur from Hudson Bay to the Mississippi River basin and the Gulf coast. On the Atlantic slope, they are known intermittently from the Delaware River to the Altamaha River, Georgia.

HABITAT AND HABITS: Quillbacks prefer large, warm, sluggish rivers and lakes and softer substrates. OCCURRENCE IN THE CHESAPEAKE BAY: Quillbacks were first described from specimens collected in the upper Chesapeake Bay, and they are resident in the larger tributaries except for the Rappahannock and York River drainages. They occur in tidal freshwater but rarely enter brackish water. They are known to tolerate salinities as great as 11‰.

REPRODUCTION: Quillbacks migrate from their deeper riverine habitats into moderate-sized streams to spawn over a wide variety of bottom types, from gravel to silt and detritus, and in a variety of flow regimes, from riffles to backwaters. Quillbacks mature at about age 4 and spawn from April to June, with females depositing up to 68,000 eggs. They may reach 11 years of age.

FOOD HABITS: Quillbacks are bottom feeders, sucking up a variety of small benthic invertebrates, such as insects, mollusks, and crustaceans, along with plant detritus and sand.

IMPORTANCE: Quillbacks are of no commercial or recreational interest.

White sucker - *Catostomus commersonii* (Lacepède, 1803)
Shorthead redhorse - *Moxostoma macrolepidotum* (Lesueur, 1817)

KEY FEATURES: White suckers have more than 50 small scales along the lateral line and a dorsal fin with a straight margin. Shorthead redhorses have fewer than 50 large scales along the lateral line and a dorsal fin with a concave margin. COLOR: White suckers are brown to olive above, grading to yellow olive on the side and white to yellowish below. Shorthead redhorses are dark olive to tan olive above, silvery to coppery on the side, and white below. SIZE: Both of these suckers have been reported to reach more than 60 cm (2 ft) TL, but 40 cm (1.3 ft) is more typical in our region. RANGE: White suckers occur from the Arctic basin to some Gulf coast drainages. On the Atlantic slope, they occur from Nova Scotia to Georgia. Shorthead redhorses occur from Canada and the Great Lakes and Mississippi drainages south to Oklahoma

White sucker *Catostomus commersonii*

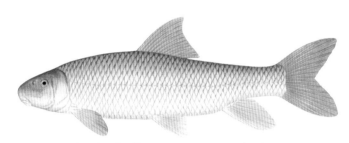

Shorthead redhorse *Moxostoma macrolepidotum*

and Alabama. On the Atlantic slope, they occur from the Hudson River to South Carolina.

HABITAT AND HABITS: White suckers occupy a wide variety of habitats, from rocky rivers to muddy creeks in the mountains and Piedmont, but they rarely occur on the Coastal Plain. Their inclusion here is based on old brackish water records from the upper Chesapeake Bay. Shorthead redhorses occur in warm rivers and streams with moderate gradient, over a variety of substrates, from mud to gravel to bedrock. They are common on the Coastal Plain and enter tidal fresh and brackish water with salinities as great as 8‰. **OCCURRENCE IN THE CHESAPEAKE BAY:** Although both these suckers are resident in all the major drainages of the Chesapeake Bay, white suckers are confined mostly to the mountains and Piedmont, while shorthead redhorses are common on the Piedmont and Coastal Plain and occur in tidal rivers, occasionally entering brackish water.

REPRODUCTION: Both these suckers spawn over gravel riffles in the spring, but white suckers tend to begin spawning earlier (March) than shorthead redhorses. White suckers mature as early as age 2 and live for as long as 17 years. Shorthead redhorses mature at 3 to 5 years of age and may live to 14 years.

FOOD HABITS: Both are benthic feeders, vacuuming up a wide variety of aquatic insects, crustaceans, mollusks, and worms, along with algae and detritus.

IMPORTANCE: Neither of these two species is of commercial or recreational value in the Chesapeake Bay region, but both may be taken recreationally, by hook and line and by gigging.

Goldfish - *Carassius auratus* (Linnaeus, 1758)

KEY FEATURES: Mouth without barbels; 25–34 large scales along lateral line; dorsal fin long, serrated spine in front of dorsal and anal fins; typically 14–21 dorsal-fin rays. COLOR: Variable, body reddish, pink, orange, gray, or black; most wild specimens are olive, brassy, or dusky gray, sometimes with blotches. SIZE: Maximum adult size 48 cm (1.6 ft) TL. RANGE: Originating in Asia and eastern Europe, this species has been introduced in freshwaters around the world. Goldfish are firmly established throughout the United States, including Hawaii.

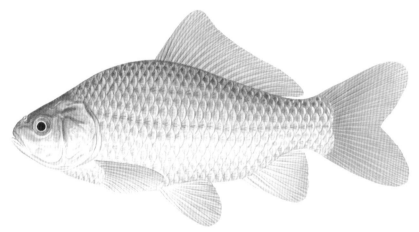

Goldfish *Carassius auratus*

HABITAT AND HABITS: Freshwater in sluggish, warm backwaters and streams as well as ponds, lakes, and reservoirs. Although able to tolerate salinities of 7–8‰ (with some reports indicating tolerance to at least 17‰), goldfish are most abundant in quiet freshwaters that support large quantities of submerged aquatic vegetation. OCCURRENCE IN THE CHESAPEAKE BAY: Introduced into Chesapeake Bay drainages as early as the 1870s, goldfish are found today primarily in the James and Potomac rivers. Occasional specimens are taken in tidal freshwater, with a rare record from near the Chesapeake Bay mouth.

REPRODUCTION: Goldfish spawn in late spring and summer over submerged vegetation or tree roots. Females spawn multiple times, depositing 2,000–4,000 adhesive eggs per spawning, with annual fecundity estimated to be as high as 400,000 eggs. Maturity may be attained as early as one year of age, and goldfish may live six to seven years in the wild.

FOOD HABITS: Goldfish are omnivorous but principally feed on phytoplankton and bottom-dwelling organisms such as insects, worms, and mollusks.

IMPORTANCE: Although goldfish are sometimes collected in pound nets near the Chesapeake Bay mouth, they are of no direct interest to commercial or recreational fisheries. However, goldfish are often sold as bait, sometimes under the name "Baltimore minnow." Their principal value is in the ornamental fish trade, where centuries of selective breeding has led to development of a myriad of varieties.

Common carp - *Cyprinus carpio* Linnaeus, 1758

KEY FEATURES: Barbel on side of snout and at rear angle of jaws; usually 35–39 scales along lateral line, but mutants called mirror carp, developed in culture, sometimes have enlarged lateral scales. Dorsal fin long, serrated spine in front of dorsal and anal fins. COLOR: Brassy olive on back to silvery on belly; scale bases dark; tail, anal, and pelvic fins may be reddish. SIZE: Maximum adult size 1.2 m (3.9 ft) TL. RANGE: Common carp are native to Asia but were introduced into Europe as early as Roman times. The species was cultured and widespread in Europe by the thirteenth century. Introduced into the United States in 1831, they are presently widespread and occur in all major watersheds.

HABITAT AND HABITS: Common carp occupy a wide range of freshwater habitats but prefer sluggish warm-water bodies and frequently occur in schools. They are most often found over soft vegetated bottoms and are tolerant of a wide range of environmental conditions, including temperatures up to 35°C (95°F) and low oxygen. Carp are capable of breathing atmospheric oxygen at the surface. OCCURRENCE IN THE CHESAPEAKE BAY: Introduced to the Chesapeake Bay region in 1877, common carp are now resident in all the major tributaries of the bay, ranging down into brackish water with salinities as high as 17.6‰.

REPRODUCTION: Common carp mature at 2 to 4 years of age and live for 9 to 15 years, rarely exceeding 20 years of age in North America. They spawn in quiet backwaters and along shallow shorelines in May and June, their activities often causing considerable splashing. Females may deposit up to two million adhesive benthic eggs during the spawning season, most often over vegetation or woody debris.

FOOD HABITS: Common carp are omnivorous, consuming aquatic and terrestrial invertebrates, small fish, plant material, and organic wastes. They most often root around on the bottom, stirring up clouds of mud, but also occasionally sip floating food items from the surface.

IMPORTANCE: Common carp are of little importance to commercial fisheries in the Chesapeake Bay region. Common carp landings in the Chesapeake Bay in 2008 were some 16.3 mt (36,000 lb), mostly from Maryland. In recreational fisheries, common carp are taken with bow and arrow, by angling with bait, and by fly fishing. Carp readily take a weighted nymph or shrimp fly and are formidable opponents on a fly rod.

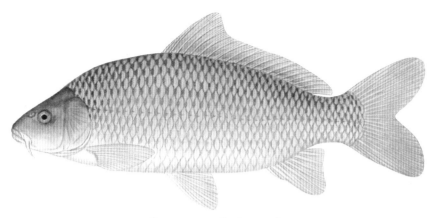

Common carp *Cyprinus carpio*

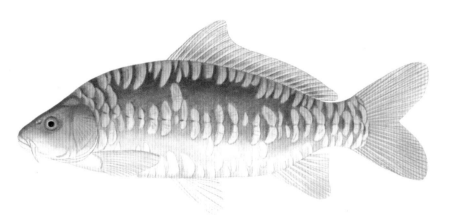

Mirror carp *Cyprinus carpio*

Eastern silvery minnow - *Hybognathus regius*
Girard, 1856

KEY FEATURES: Dorsal fin short-based with 7 to 10 (typically 8) rays; 8 to 9 (typically 8) anal-fin rays; peritoneum is dark and intestine is long and coiled like a watch spring. **COLOR:** Olive on back, silvery on sides, with some pigmentation peppered below lateral line. **SIZE:** Maximum adult size 15 cm (5.9 in) TL. **RANGE:** Atlantic slope from the Lake Ontario / St. Lawrence River basin to the Altamaha River in Georgia.

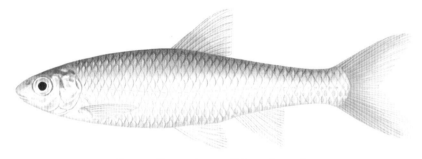

Eastern silvery minnow *Hybognathus regius*

HABITAT AND HABITS: Eastern silvery minnows mostly occur in large streams and sluggish rivers, usually in backwaters over soft bottom. They are frequently found in tidal fresh and slightly brackish water. Eastern silvery minnows live in schools near the bottom, often in association with other minnows and shiners. **OCCURRENCE IN THE CHESAPEAKE BAY:** Eastern silvery minnows are resident in all tributaries of the bay and have been taken in salinities as high as 14‰.

REPRODUCTION: Eastern silvery minnows mature in two years and live for three years. Spawning occurs in April and May, over silt to sand and fine gravel bottoms, in shallow coves and tidal creeks. Vigorous vibrations during spawning often stir up the bottom and muddy the water. Females may deposit 2,000–6,000 benthic eggs during the spawning season.

FOOD HABITS: Eastern silvery minnows' long, coiled gut is an adaptation to consume plant material, benthic diatoms, algae, and detritus.

IMPORTANCE: Eastern silvery minnows have no importance to recreational or commercial fisheries. They are an important forage fish where abundant.

Golden shiner - *Notemigonus crysoleucas* (Mitchill, 1814)

KEY FEATURES: Lateral line curves down behind pectoral fin, 39–57 scales in lateral line; dorsal fin with 7–9 rays; anal fin with 8–19 rays (typically 10–15) and falcate margin. COLOR: Brassy to silvery on sides, back darker; young have dark stripe along sides; fins yellowish to reddish in breeding adults. SIZE: Maximum adult size 31 cm (1 ft) TL. RANGE: Native to eastern and central North America from Canada to Florida and Mexico. Because of its use as a baitfish, golden shiners have been widely introduced throughout North America, including west of the Rocky Mountains.

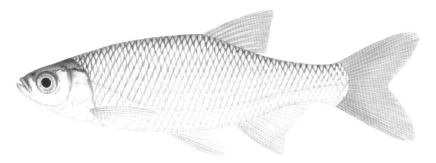

Golden shiner *Notemigonus crysoleucas*

HABITAT AND HABITS: Golden shiners occupy a wide variety of warm-water, sluggish habitats in streams, rivers, swamps, and lakes. They can tolerate low oxygen levels and high turbidity and temperature. Usually found in midwater or near-surface waters in small, loosely aggregated schools. OCCURRENCE IN THE CHESAPEAKE BAY: Common to abundant in all tributaries of the Chesapeake Bay. Occasionally enter brackish areas with salinities as high as 17‰.

REPRODUCTION: Most golden shiners mature at age two and may live for nine years. Spawning occurs over vegetation or gravel from April to August (or later). Females may deposit up to 200,000 (or more) adhesive benthic eggs during the spawning season.

FOOD HABITS: Golden shiners are principally pelagic planktivores and consume small crustaceans and insects. They also may forage on benthic insects, snails, and even algae.

IMPORTANCE: Golden shiners are of no direct importance to commercial or recreational fisheries. However, they have been widely cultured for use as a forage fish in bass ponds and for live bait by recreational fishers.

Spotfin shiner - *Cyprinella spiloptera* (Cope, 1867)
Bridle shiner - *Notropis bifrenatus* (Cope, 1867)
Spottail shiner - *Notropis hudsonius* (Clinton, 1824)

KEY FEATURES: These three species can best be separated by their color patterns. COLOR: Spotfin shiners usually lack lateral stripe (may be present on rear of body in juveniles) and have dusky pigmentation on rear of dorsal fin. Bridle shiners may be distinguished by black lateral stripe that continues from tip of snout to base of tail. Spottail shiners have dark lateral stripe that begins behind gill cover and usually becomes more prominent, ending as dark spot toward base of tail. SIZE: Bridle shiners are small, 6 cm (2.4 in) TL. Spotfin shiners are a bit larger, 11 cm (4.3 in) TL, but more typically 6–9 cm (2.4–3.5 in) TL. Spottail shiners are the largest of the three, attaining 15 cm (5.9 in) TL. RANGE: Bridle shiners are confined to Atlantic slope drainages from Maine to South Carolina, but their distribution, particularly in the South, is discontinuous. Spottail shiners are one of the most widely distributed freshwater fishes in North America, extending from the Mackenzie River in northern Canada throughout the Great Lakes and Mississippi basin and along the Atlantic slope from the St. Lawrence River to Georgia. Spotfin shiners are found in the Mississippi basin, in the Great Lakes, and along the Atlantic slope from the Hudson River to the Potomac.

HABITAT AND HABITS: Although the three species are known to occupy a variety of habitats from large rivers and smaller streams to lakes, bridle shiners prefer shallower vegetated areas, and the other two prefer deeper water in larger rivers. Spotfin shiners seem to prefer areas with more current than the other two species. Of the three, spottail shiners are most likely to occur regularly in brackish areas. OCCURRENCE IN THE CHESAPEAKE BAY: Bridle shiners were once common in most tributaries of the bay but have become rare or extirpated over much of their former range. They seem to be persisting in the James and Chickahominy Rivers (but strangely, not the Appomattox). The mysterious decline of bridle shiners extends over their entire range and may be associated with habitat alteration, increased turbidity, and loss of submerged aquatic vegetation. Spottail shiners are common in all bay tributaries and, of the three species, are the only one to occur on a regular basis in brackish water. Spotfin shiners occur in bay tributaries from the Potomac River north.

REPRODUCTION: Bridle shiners mature in 1 year and deposit 1,000–2,000 benthic adhesive eggs over submerged vegetation during their early summer spawning season. Spottail shiners mature at one to three years and deposit benthic adhesive eggs over sand or gravel in shallow areas with current. Spotfin shiners mature at 1–3 years

Spotfin shiner *Cyprinella spiloptera*

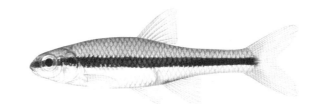

Bridle shiner *Notropis bifrenatus*

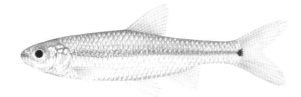

Spottail shiner *Notropis hudsonius*

and deposit up to 7,500 benthic adhesive eggs in crevices of submerged woody debris.

FOOD HABITS: All three species feed on small crustaceans, aquatic insects, and plant material. Bridle and spottail shiners are more likely to feed on the bottom and consume other benthic food organisms, whereas spotfin shiners feed more often in midwater.

IMPORTANCE: These small species have no importance to recreational or commercial fisheries, although larger spottail shiners are sometimes used as bait. All are important forage species where abundant.

Gafftopsail catfish - *Bagre marinus* (Mitchill, 1815)

KEY FEATURES: Two pairs of well-developed elongate barbels around mouth, one pair on upper and one on lower jaw, barbels on upper jaw noticeably flattened and very long, extending to rear of first dorsal fin; dorsal-fin spine with long filament, frequently reaching to adipose fin; caudal fin deeply forked, its upper lobe slightly longer; pectoral-fin spine with long filament reaching to anal-fin origin. COLOR: Bluish gray to brown above, silvery on sides, and white below. SIZE: Maximum adult size 69 cm (2.3 ft) TL, but typically not greater than 50 cm (1.6 ft) TL. RANGE: Western North Atlantic from Cape Cod (rare) to Brazil, including the Gulf of Mexico and Caribbean Sea.

Gafftopsail catfish *Bagre marinus*

HABITAT AND HABITS: Gafftopsail catfish are coastal and enter higher-salinity areas of estuaries over soft substrates. They prefer subtropical and tropical latitudes but extend their range in summer to warm-temperate regions. OCCURRENCE IN THE CHESAPEAKE BAY: Gafftopsail catfish are rare summer visitors to the lower Chesapeake Bay and have been reported as far north as the Potomac River.

REPRODUCTION: Gafftopsail catfish spawn in May and June. The males hold the eggs in their mouth for up to two months before they hatch. Males cease feeding during the incubation period.

FOOD HABITS: Gafftopsail catfish are opportunistic bottom feeders, taking a variety of invertebrates and small fishes.

IMPORTANCE: Due to their low abundance in the Chesapeake Bay region, the gafftopsail catfish is not targeted in either commercial or recreational fisheries. Their sharp spines can produce painful puncture wounds, so this species should be handled with care.

White catfish - *Ameiurus catus* (Linnaeus, 1758)

KEY FEATURES: Tail moderately forked, with rounded margin; anal fin rounded, its base much longer than its height, with 21–26 rays (typically 22–24). **COLOR:** Gray to bluish gray dorsally, whitish ventrally; chin barbels pale, other barbels dusky. **SIZE:** Maximum adult size 60 cm (2 ft) TL. Typical size is 30–33 cm (12–13 in) TL. Often confused with channel catfish; however, white catfish rarely exceed 2.7 kg (6 lb), whereas channel catfish reach 34 kg (75 lb). **RANGE:** Atlantic slope from the Delaware River to Florida and the eastern drainages of the Gulf of Mexico. Cultured and widely introduced elsewhere.

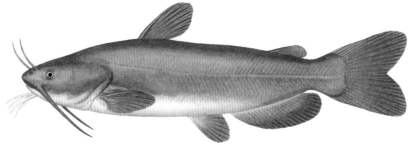

White catfish *Ameiurus catus*

HABITAT AND HABITS: White catfish occur in lakes of various sizes and medium to large warm sluggish rivers. Common in tidal fresh and brackish water to 10‰ and occasionally to 27‰. White catfish are tolerant of natural acid water conditions and occur in swamp creeks and still waters such as Lake Drummond in the Dismal Swamp, Virginia. **OCCURRENCE IN THE CHESAPEAKE BAY:** White catfish are abundant in all the major tributaries of the bay, in creeks of moderate size to rivers.

REPRODUCTION: White catfish mature at 3 to 4 years and have a maximum life span of 14 years. Spawning occurs from late May to July. A spawning pair clears a nest almost a meter (3.3 ft) in diameter, in sand or gravel, in which the female may deposit 1,400–3,500 eggs. One or both parents protect the eggs and developing brood.

FOOD HABITS: Young white catfish principally consume insects but become more omnivorous with age. Older white catfish eat a variety of benthic invertebrates, fishes, and plant material.

IMPORTANCE: White catfish are taken along with other freshwater catfishes in both recreational and commercial fisheries. The latter employs mostly pots and fyke nets but also virtually all other gear types legal in the Chesapeake Bay. Commercial landings from the bay region in 2008 for all catfish species combined (except blue catfish) were 900 mt (2 million lb). White catfish have undergone significant declines in bay tributaries over the last decade, apparently due to competition with the introduced blue catfish. Anglers catch catfishes fishing on the bottom with a variety of baits, including worms, live minnows, cut bait, chicken livers, and "stinkbait," the preparation of which is a closely guarded secret among some devotees.

Yellow bullhead - *Ameiurus natalis* (Lesueur, 1819)

Brown bullhead - *Ameiurus nebulosus* (Lesueur, 1819)

KEY FEATURES: These two species are distinguished by the color of the chin barbels. COLOR: Yellow bullheads are olive brown above and along the side, grading to yellowish white below; chin barbels pale. Brown bullheads are black to gray brown to dirty white below, sometimes with mottling on the side; chin barbels dusky. SIZE: Both these bullheads may reach a maximum size of about 47 cm (1.5 ft) TL, but the average size is about 18–23 cm (7–9 in) TL. RANGE: Both these species are widely distributed from southern Canada, the Great Lakes, and throughout the Mississippi basin to the Gulf of Mexico, and along the Atlantic slope from Florida north to New York (yellow bullhead) and Nova Scotia (brown bullhead).

HABITAT AND HABITS: Both of these bullheads frequent ponds, lakes, and warm sluggish creeks and rivers. Both species are tolerant of acidic water, but yellow bullheads are the more tolerant of the two. Conversely, brown bullheads are more tolerant of brackish water (13‰) than yellow bullheads and are far more likely to be encountered in an estuary. OCCURRENCE IN THE CHESAPEAKE BAY: Both occur in many Chesapeake Bay tributaries, from small creeks to large rivers. Brown bullheads are much more common in brackish water than yellow bullheads.

REPRODUCTION: Yellow bullheads mature at age two and may live for at least six years. Brown bullheads mature at age three and may live for nine years. Both species spawn from April into mid-summer. Spawning pairs excavate a nest on sandy bottom in shallow water, often near cover. Females deposit 200–700 eggs, which are guarded by both parents.

Yellow bullhead *Ameiurus natalis*

Brown bullhead *Ameiurus nebulosus*

FOOD HABITS: Both species are omnivorous and consume a spectrum of benthic invertebrates (including insects) and small fishes. Brown bullheads may consume more algae than yellow bullheads. Both are known to scavenge.

IMPORTANCE: Bullheads are taken along with our other freshwater catfishes in both recreational and commercial fisheries.

Blue catfish - *Ictalurus furcatus* (Valenciennes, 1840)

KEY FEATURES: Tail deeply forked; anal fin with 27–38 rays (typically 30–36) and straight distal edge. COLOR: Back gray to blue gray, sides silvery, belly white; no spots at any size. SIZE: Maximum adult size 1.7 m (5.6 ft) TL. Typical adult size is less than 60 cm (2 ft) TL. A 1-meter (39-in) TL individual can weigh more than 22.5 kg (50 lb). The largest catfish ever recorded in the United States was a 68-kg (150-lb) blue catfish, which was weighed in the late nineteenth century but never measured. However, there are anecdotal reports of larger specimens of blue catfish. RANGE: Native to the Mississippi River basin and tributaries of the Gulf of Mexico. Widely introduced on the Atlantic coast.

HABITAT AND HABITS: Blue catfish occur in larger rivers and creeks as well as in impoundments. They penetrate downstream in tidal rivers to at least 14.7‰. OCCURRENCE IN THE CHESAPEAKE BAY: Blue catfish were first introduced by the Virginia Department of Game and Inland Fisheries into the Rappahannock River in 1974 and into the James River in 1975. Subsequently, the species has spread and is abundant in all major Chesapeake Bay tributaries.

REPRODUCTION: Blue catfish mature at about 4 years of age and live to at least 21 years. They spawn in sheltered nests, under logs and undercut banks, in late spring and early summer. Both sexes share in brooding. Female fecundity is about 2,000 eggs per 0.45 kg (1 lb) body weight. Thus, a 41-kg (90-lb) female would be expected to lay 180,000 eggs.

FOOD HABITS: Young blue catfish feed mostly on small aquatic invertebrates. Larger fish feed on a wide variety of invertebrates, including clams as well as fishes. Very large blue catfish are largely piscivorous.

IMPORTANCE: Blue catfish are the target of both recreational and commercial fisheries. Recreational fishermen have taken trophy blue catfish in excess of 45.5 kg (100 lb) from bay tributaries. After their introduction, blue catfish taken in commercial fisheries were simply recorded in the fishery statistics with all the other catfishes taken in the bay. However, by 2003, blue catfish had spread and become so abundant as to warrant a separate category in the statistics. In 2008, the combined landings of blue catfish from Maryland and Virginia were 167.6 mt (368,780 lb). Regardless of the blue catfish contribution to both recreational and commercial fisheries in Chesapeake Bay, the ecological impact of this alien species to the system has been considerable. As blue catfish increased in abundance, native white catfish declined, probably due to competition. In addition, the predatory impact of this large piscivore

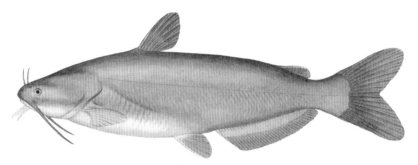

Blue catfish *Ictalurus furcatus*

in the heart of the nursery grounds for striped bass, river herrings, and shads has to be questioned.

Channel catfish - *Ictalurus punctatus* (Rafinesque, 1818)

KEY FEATURES: Tail deeply forked, caudal-fin lobes mostly straight; anal fin long, with distinctly rounded bottom edge, 23–32 (typically 26–27) anal-fin rays. COLOR: Blue or gray to greenish gray dorsally, whitish ventrally; chin barbels white to dusky; body usually with dark gray to black spots, larger individuals sometimes lose spots. SIZE: Maximum adult size 121 cm (4 ft) TL. Typical adult size is less than 60 cm (2 ft) TL. RANGE: Native to the Hudson Bay, Great Lakes, Mississippi River, and Gulf of Mexico drainages. Introduced widely along the Atlantic slope and elsewhere.

Channel catfish *Ictalurus punctatus*

HABITAT AND HABITS: Although channel catfish occur in lakes and ponds, they are more often encountered in large streams or rivers. Adult channel catfish occupy deep pools in and around logs and other cover. Juveniles occur in faster-moving, shallower waters. OCCURRENCE IN THE CHESAPEAKE BAY: Channel catfish were introduced into the mid-Atlantic region and now are common in all tributaries of the Chesapeake Bay. They have been collected in waters with salinities as great as 20‰ and are often encountered in waters of 5‰ or greater.

REPRODUCTION: Channel catfish may mature as early as age 2 but usually do so at age 4, and they may live for about 22 years. Spawning extends from mid-May to early July. Females usually deposit 4,200–10,600 eggs in a sheltered nest subsequently guarded by the males. Channel catfish can live as long as 14 years, with some reported to exceed 25 years.

FOOD HABITS: Young channel catfish eat plankton and aquatic insect larvae, whereas larger fish consume a variety of invertebrates, plant material, and fishes. The proportion of fish in the diet increases with age and size.

IMPORTANCE: Channel catfish are taken in large numbers in both recreational and commercial fisheries in the Chesapeake region and elsewhere. Channel catfish are widely available for sale in supermarkets, as the firm-bodied white flesh is quite palatable.

Margined madtom - *Noturus insignis* (Richardson, 1836)

KEY FEATURES: Head flattened, with four barbels (whiskers) on each side around mouth; dorsal and pectoral fin with pungent spine in front; tail roundish or squarish, not forked; small fleshy adipose fin (behind dorsal fin) attached to tail fin. COLOR: Back and sides brown to olive gray; belly white; median fins pale, with dusky margin; barbels dark on snout, white on chin. SIZE: Madtoms are the smallest members of the North American catfish family. The maximum adult size of margined madtoms is 17.9 cm (7 in) TL; common to 10 cm (3.9 in) TL. RANGE: Margined madtoms occur in Atlantic slope drainages from the Hudson (New York) to the Altamaha River (Georgia) and in the New River in Virginia. They have been widely introduced elsewhere.

HABITAT AND HABITS: Margined madtoms are found in larger streams and rivers in riffles, runs, and pools and are widely distributed from the Appalachian Ridge and Valley and from the Blue Ridge to the Piedmont and Coastal Plain. They seek cover during the day and actively forage at night. OCCURRENCE IN THE CHESAPEAKE BAY:

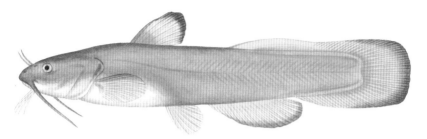

Margined madtom *Noturus insignis*

Common in Coastal Plain streams and rivers and occasionally brackish water up to 5‰.

REPRODUCTION: Margined madtoms mature in one to two years and live at least four years. Spawning occurs in May, in nests under flat rocks situated in slow runs. Females deposit 50 to more than 200 eggs, which are guarded by the males.

FOOD HABITS: Margined madtoms mainly eat aquatic insects on or near stream or river bottoms. They are nocturnal feeders.

IMPORTANCE: Margined madtoms are not harvested by recreational or commercial fisheries but are widely used as a preferred live bait by bass fishers. They should be handled carefully, because the venomous spines can inflict painful puncture wounds.

Flathead catfish - *Pylodictis olivaris* (Rafinesque, 1818)

KEY FEATURES: Head very flattened; lower jaw projecting beyond upper; tail rounded, not forked, posterior margin of caudal fin slightly notched. COLOR: Yellow to olive and brown above, mottled brown on sides, yellow white below. SIZE: Maximum adult size 1.4 m (4.6 ft) TL with weight exceeding 55 kg (120 lb), but usually 40–90 cm (1.3–2.6 ft) TL. RANGE: Native to the southern Great Lakes and Mississippi River and Gulf Coast drainages. Widely introduced elsewhere.

Flathead catfish *Pylodictis olivaris*

HABITAT AND HABITS: Flathead catfish favor lakes and large warm rivers and streams. Flathead catfish usually live alone and lie in depressions during the day in deeper water or under logs and banks, emerging to forage in shallower areas at night. The young occur under rocks in riffles. These catfish rarely enter brackish water. OCCURRENCE IN THE CHESAPEAKE BAY: Flathead catfish were introduced into the lower Potomac River drainage in 1963 and into the lower James River sometime between 1965 and 1977. They appear to be limited to these two drainages, probably because of their aversion to brackish water, and are not common in either watershed.

REPRODUCTION: Flathead catfish mature in 3 to 5 years and may reach 19 years of age. Spawning occurs in June and July. Spawning pairs clear out a nest on the bottom, usually near cover. Females may deposit 6,900–11,300 eggs.

FOOD HABITS: Young flathead catfish consume small benthic crustaceans and insect larvae, whereas larger, older individuals consume crayfish, clams, and fishes. Adults move at night from deeper water or cover to riffles and shallows to feed.

IMPORTANCE: Because of their low abundance in the Chesapeake Bay region, there are no directed fisheries for flathead catfish, but they occur in landings of fisheries focused on other catfish. They are occasionally taken by anglers in the blue catfish trophy fishery on the James River near Hopewell, Virginia.

Redfin pickerel - *Esox americanus* Gmelin, 1789

KEY FEATURES: Snout long, depressed, and broad (snout shorter than in chain pickerel); distance from snout tip to rear of jaws about 8–9 times in TL. **COLOR:** Dark green to black above, with series of 20–36 dark wavy vertical bars along sides, and white below; dark oblique bar slanting backward beneath eye; lower fins reddish. **SIZE:** Maximum adult size 37 cm (1.2 ft) TL. **RANGE:** Redfin pickerel (*Esox americanus americanus*) occur on the Atlantic slope from the St. Lawrence Seaway to Florida. In Florida they intergrade with grass pickerel (*E. americanus vermiculatus*), which occur primarily from the southern Great Lakes and Mississippi drainage to the Gulf slope.

Redfin pickerel *Esox americanus*

HABITAT AND HABITS: Redfin pickerel are found in small coastal plain streams, lakes, ponds, and tannin-stained swamps, where they are associated with aquatic vegetation. Redfin pickerel often occur with chain pickerel; when in ponds, chain pickerel are usually more abundant than redfin pickerel, while in small streams, the opposite is true. The life history of redfin pickerel is similar in many respects to that of chain pickerel. **OCCURRENCE IN THE CHESAPEAKE BAY:** Redfin pickerel are common in small tributaries of the Chesapeake Bay and have been taken in tidal waters with salinities as high as 10‰.

REPRODUCTION: Redfin pickerel mature in two to three years and reach a maximum age of six. They spawn in February and March over heavily vegetated pond or stream margins, the female depositing 180–550 eggs. There is no parental care.

FOOD HABITS: Young redfin pickerel prey on invertebrates and aquatic insects but become more piscivorous with age. They are diurnal feeders.

IMPORTANCE: Redfin pickerel are not commercially important and, due to their small size, are not regarded as sport fish. However, they will readily strike at lures and live bait.

Chain pickerel - *Esox niger* Lesueur, 1818

KEY FEATURES: Snout long, depressed, and broad (snout longer than redfin pickerel); distance from snout tip to rear of jaws less than 7 times in TL. **COLOR:** Juveniles dark green to black above, with series of dark wavy vertical bars along sides, and white below; dark vertical bar beneath eye; lower fins pale. Specimens larger than about 30.5 cm (12 in) TL with dark chain-like pattern along sides. **SIZE:** Maximum adult size 99 cm (3.2 ft) TL but usually less than 75 cm (2.5 ft) TL. **RANGE:** Atlantic slope from New England to Florida, Gulf slope, and lower Mississippi valley.

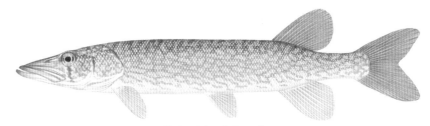

Chain pickerel *Esox niger*

HABITAT AND HABITS: Chain pickerel occur on the Coastal Plain and Piedmont in shallow vegetated areas of sluggish rivers, streams, ponds, and lakes and are very tolerant of naturally acidic waters. They are usually solitary, hanging out near submerged structures, waiting to ambush passing prey. **OCCURRENCE IN THE CHESAPEAKE BAY:** Chain pickerel are abundant in most Chesapeake Bay tributaries and in tidal waters. Although more often encountered in freshwaters, chain pickerel can tolerate salinities up to 22‰.

REPRODUCTION: Chain pickerel mature in two to four years and may reach a maximum age of nine years. Chain pickerel spawn in February and March over shallow weed beds, females depositing 340–8,000 or more eggs in strings. There is no parental protection.

FOOD HABITS: Young chain pickerel feed on various aquatic invertebrates but soon begin to concentrate on fishes, including smaller members of their own species. Larger chain pickerel are voracious piscivores but may also include frogs, snakes, and mice in their diet. Their ambush style of hunting requires cover, usually sought in the form of aquatic plants, tree stumps, and fallen logs.

IMPORTANCE: Chain pickerel are a popular sport fish because of their fighting qualities and proclivity to take both live and artificial baits. However, angling for this species presents difficulties, as chain pickerel are typically found in waters clogged with weeds and brush. Live minnows, spoons, spinners, and crank baits are all enthusiasti-cally attacked, usually with a violent tug followed by an aerial display. The battle is usually short-lived, because chain pickerel are adapted to accelerate rapidly but not to maintain high swimming speeds for more than a brief time. Larger chain pickerel are good to eat, with flaky white flesh but a myriad of small bones.

Eastern mudminnow - *Umbra pygmaea* (DeKay, 1842)

KEY FEATURES: Dorsal fin originates nearer caudal fin than head, spineless, low, and rounded posteriorly; caudal fin rounded; anal fin originates on a vertical with midpoint of dorsal fin, anal fin low and rounded posteriorly. **COLOR:** Dark brown on back, lighter brown on sides and belly; flanks with 10–12 horizontal stripes; dark spot or bar at base of tail. **SIZE:** Maximum adult size 15.2 cm (6 in) TL. Commonly 7.5 cm (3 in) or less TL. **RANGE:** Atlantic slope from southern New York to northern Florida, where they are also found in some Gulf slope drainages.

Eastern mudminnow *Umbra pygmaea*

HABITAT AND HABITS: The common name for the family is derived from their habit of burying themselves in the bottom substrate, typically mud. Some mudminnows are reported to be resistant to freezing and able to overwinter in shallow frozen water bodies. Eastern mudminnows are typically swamp dwellers and also occur in sluggish streams and ponds, hiding in aquatic vegetation by day and foraging at night. Mud-minnows typically rise to the surface periodically to gulp air, particularly when water temperatures rise or oxygen values become low. **OCCURRENCE IN THE CHESAPEAKE BAY:** Common in low-lying streams and swamps; occasionally in tidal waters with salinities as high as 17‰.

REPRODUCTION: Eastern mudminnows mature at one or two years and live for three years. Spawning takes place in March and April in backwaters, in nests constructed in masses of vegetation or on the bottom near vegetation. Females may deposit from about 100 to more than 2,500 eggs. One or both parents may guard the nest.

FOOD HABITS: Eastern mudminnows consume small larval insects, crustaceans, and snails.

IMPORTANCE: Mudminnows have no importance to recreational or commercial fisheries. Due to their hardiness, mudminnows are sometimes used as baitfish.

Inshore lizardfish - *Synodus foetens* (Linnaeus, 1766)

KEY FEATURES: Large mouth with sharp teeth on jaws and tongue; short dorsal fin located over middle of body, followed by small adipose fin in front of forked tail; pelvic fins in front of vent and larger than pectoral fins. COLOR: Greenish brown on back, sides white with dark diamond-shaped blotches, belly white. Inshore lizardfish are apparently able to assume various patterns depending on conditions of background and light. SIZE: Maximum adult size 45 cm (1.5 ft) TL. RANGE: Coastal waters from Cape Cod to Brazil.

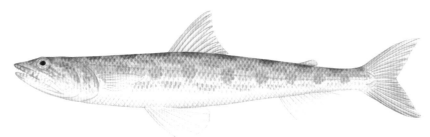

Inshore lizardfish *Synodus foetens*

HABITAT AND HABITS: Inshore lizardfish are a subtropical/tropical species that enters warm-temperate waters in summer. They are found on both shallow and deep sand flats among vegetation, in the tidal portions of large rivers, and in deep channels. They are probably more common over mud than shell bottoms. Lizardfish are known to partially bury themselves in sandy or muddy substrates and conceal themselves from unsuspecting prey. Although the common name is inshore lizardfish, *S. foetens* has been collected from depths of up to 200 m (660 ft). OCCURRENCE IN THE CHESAPEAKE BAY: Inshore lizardfish are frequent summer visitors to the lower and middle Chesapeake Bay, extending as far north as the Chester River. They may penetrate tributaries upstream occasionally to salinities lower than 2‰ but are more likely encountered at salinities above 20‰.

REPRODUCTION: Inshore lizardfish spawn in summer on the continental shelf, primarily south of Cape Hatteras. Little is known about age, growth, or reproduction in this species.

FOOD HABITS: Inshore lizardfish are voracious lie-in-wait predators that can dart quickly to capture small fishes. While inshore lizardfish are mostly piscivorous, they will also include shrimps, crabs, and cephalopods in their diet.

IMPORTANCE: Because of their small size, inshore lizardfish are of no importance to commercial or recreational fisheries. However, they are frequently caught and discarded by sport fishers because of their proclivity to take actively moving artificial lures or even cut bait.

Striped cusk-eel - *Ophidion marginatum* DeKay, 1842

KEY FEATURES: Eel-like body with long continuous dorsal, tail, and anal fin running around center margin of body; pelvic-fin rays reduced and appear as two branched filaments in throat region. **COLOR:** Back grayish green, sides golden, belly white; two or three dark or dusky stripes along sides. **SIZE:** Maximum adult size 25 cm (10 in) TL. **RANGE:** Southern New England to northern Florida.

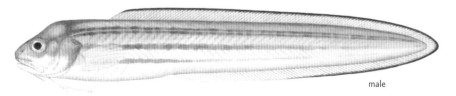

male

Striped cusk-eel *Ophidion marginatum*

HABITAT AND HABITS: Striped cusk-eels are an elusive species that is probably much more common than records would indicate. They burrow tail-first into soft substrates, where they remain hidden during the day, emerging at night to forage. The species has been recorded from estuaries and the nearshore coastal zone from spring to autumn. However, its whereabouts in winter are a mystery, and it may move offshore or simply remain burrowed and inactive in the colder months. The swim bladder, anterior vertebrae, and dorsal cranial muscles are highly modified to form a drumming apparatus, with the details of this apparatus differing between the sexes. OCCURRENCE IN THE CHESAPEAKE BAY: Striped cusk-eels have been reported occasionally from the lower bay, but they are probably regular cryptic summer residents.

REPRODUCTION: Based on lengths of mature individuals and monthly growth increments, striped cusk-eels probably mature at one year of age, and they may live for three or four years. Spawning occurs inshore near estuarine inlets in summer, at night. Courtship involves sound production and close, if brief, contact between the spawning pair, after which the female releases a buoyant gelatinous sac in which a small batch of eggs are enclosed. Individual females may spawn nightly for as long as two months. Late juvenile and adult males are distinguished from females by the presence of a hump on top of the head posterior to the eyes, formed by the hypertrophy of bone and muscles associated with the sound-producing apparatus.

FOOD HABITS: Striped cusk-eels consume crustaceans and small fishes.

IMPORTANCE: Striped cusk-eels have no recreational or commercial fisheries importance.

Atlantic cod - *Gadus morhua* Linnaeus, 1758

KEY FEATURES: Snout round, projecting over lower jaw; prominent chin barbel, about equal to eye diameter; teeth small, pointed, and in bands on jaws; three dorsal fins; first dorsal fin the highest, with convex outer margin; second and third dorsal fins tapering posteriorly, with outer margins nearly straight; two anal fins, first anal fin longer than second. COLOR: Back gray green to black to reddish brown, paler on sides, shading to white on belly; sides covered with small round spots. SIZE: Maximum adult size 2 m (6.6 ft) TL. Most specimens are less than 1 m (3.3 ft) TL. RANGE: Atlantic cod occur on both sides of the North Atlantic. In the western North Atlantic they are found from Greenland to North Carolina (winter).

HABITAT AND HABITS: Atlantic cod are coastal cold-water bottom fish with temperature preferences of 0–10°C (32–50°F). In the mid-Atlantic area they migrate from New England in late winter and early spring as least as far south as the New Jersey coast. Historically (in the early twentieth century) when winters were colder, Atlantic cod occurred as far south as Cape Hatteras. OCCURRENCE IN THE CHESAPEAKE BAY: Atlantic cod used to occur sporadically in the first half of the twentieth century in pound-net catches at the bay mouth in March. There are no recent records.

Atlantic cod *Gadus morhua*

REPRODUCTION: Atlantic cod mature at 2 to 4 years of age and may live for more than 20 years. Spawning occurs over the continental shelf in winter and early spring. A large (1-m, or 3.3-ft) female may produce 4–8 million pelagic eggs a year.

FOOD HABITS: Young Atlantic cod feed mostly on small crustaceans but switch to a diet of mainly fishes, crabs, and squids as they grow. Herring, sand lance, silver hake, and capelin are among the most frequently eaten fish prey.

IMPORTANCE: Of no commercial or recreational interest in the Chesapeake Bay region, due to their scarcity.

Pollock - *Pollachius virens* (Linnaeus, 1758)

KEY FEATURES: Snout pointed, lower jaw projects beyond snout; tiny barbel at tip of lower jaw in juveniles, absent in adults; three dorsal fins and two anal fins; first anal fin much longer than second dorsal fin. **COLOR:** Olive green above, shading to light gray on belly; lateral line pale; caudal and dorsal fins dusky green. **SIZE:** Maximum adult size 1.3 m (4.3 ft) TL, but usually 1.1 m (3.5 ft) TL or less. **RANGE:** Pollock occur on both sides of the North Atlantic. In the western North Atlantic they are found from Newfoundland to North Carolina (winter).

Pollock *Pollachius virens*

HABITAT AND HABITS: Pollock are cold-temperate coastal fish that undertake both north/south and onshore/offshore migrations in response to seasonal temperature changes. They are active swimmers, occupying any and all levels between the surface and the bottom, and occur in inshore and offshore waters to about 200 m (660 ft) deep. **OCCURRENCE IN THE CHESAPEAKE BAY:** Pollock are rare to occasional visitors to the lower Chesapeake Bay in late winter and early spring.

REPRODUCTION: Pollock mature at 2 to 3 years of age and may live for 19 years. Spawning occurs off the New England coast in winter. Estimates of egg production are as high as four million per female.

FOOD HABITS: Pollock are voracious predators. Young pollock feed mostly on small krill, whereas older individuals consume larger krill, fishes, and squids.

IMPORTANCE: Of no commercial or recreational interest in the Chesapeake Bay region, due to their scarcity.

Silver hake - *Merluccius bilinearis* (Mitchill, 1814)

KEY FEATURES: Large mouth with prominent rows of sharp teeth and underslung lower jaw; chin barbel absent; first dorsal fin short-based and triangular, second dorsal fin long and notched mid-length. **COLOR:** Dusky gray or brownish dorsally, flanks and belly silvery and highly iridescent. **SIZE:** Maximum adult size 76 cm (2.5 ft) TL. **RANGE:** Atlantic coast from Newfoundland to South Carolina, occasionally to Florida.

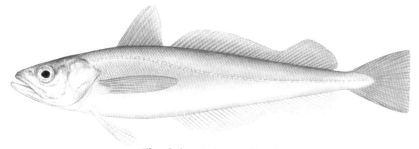

Silver hake *Merluccius bilinearis*

HABITAT AND HABITS: Silver hake are a cold-temperate coastal species that undertake seasonal migrations to stay within their preferred water temperature range (7–10°C, or 45–50°F). They are swift, strong swimmers. Silver hake migrate into our area in the winter and spring, from offshore and from the north. Part of the population is resident on the outer continental shelf and upper slope, where silver hake may form large aggregations. This species spends the day close to the bottom and rises up into the water column to feed at night. **OCCURRENCE IN THE CHESAPEAKE BAY:** Silver hake are regular visitors to the lower bay from late autumn to early spring and occur occasionally in salinities as low as 13‰. They are reported as far north in the Chesapeake Bay as the Chester River.

REPRODUCTION: Silver hake mature in 2 to 3 years and may live for 13. Spawning occurs offshore in the summer, and a single 3-year-old female may spawn 3 times, producing a total of 300,000–400,000 eggs per season. The eggs and larvae are pelagic.

FOOD HABITS: Silver hake are voracious predators living in large schools, sometimes at considerable depths. Young silver hake feed on small shrimp and other crustaceans, whereas older hake, particularly females, feed more on fishes and squids.

IMPORTANCE: A valuable commercial species in the northern part of the range, collected by trawls and trap nets.

Red hake - *Urophycis chuss* (Walbaum, 1792)

KEY FEATURES: First dorsal fin with filamentous ray; two pelvic-fin rays, filamentous and joined at base, the longer of which reaches beyond the vent. **COLOR:** Dark brown to reddish brown on back, mottled on sides, and white below, sometimes with yellowish cast; dark blotch on gill cover; first dorsal fin brown. **SIZE:** Maximum adult size about 52 cm (1.7 ft) TL. **RANGE:** Along the western Atlantic coast from Nova Scotia to North Carolina.

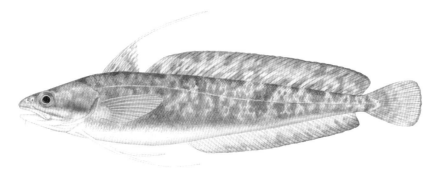

Red hake *Urophycis chuss*

HABITAT AND HABITS: Red hake are a cold-temperate coastal species that undertakes onshore-offshore seasonal migrations to remain in their preferred temperature range of 5–12°C (41–54°F). From late spring until early summer, red hake move from deep to shallow waters. As waters warm during the summer, red hake migrate to deeper water offshore (as deep as 550 m, or 1788 ft) and stay offshore until the following spring. After descending from the plankton, small juveniles become resident within the shells of live sea scallops on the continental shelf for a few months in autumn and winter. Red hake are demersal and prefer mud or sand bottoms. Juveniles live along coasts at shallow depths (4–6 m, or 13–20 ft), while larger, older red hake dwell in waters deeper than 35 m (114 ft) and may become resident on the outer continental shelf and slope. **OCCURRENCE IN THE CHESAPEAKE BAY:** Juvenile red hake are regular visitors to the lower Chesapeake Bay in late winter and spring. They can tolerate salinities as low as 21‰ and occasionally may move into the mid-bay, extending as far north as the Patuxent River.

REPRODUCTION: Red hake mature at about age 1 or 2 and may reach 14 years of age. The protracted spawning season peaks in July, and spawning occurs off southern New England and all around the inshore areas of the Gulf of Maine. The eggs and larvae are pelagic.

FOOD HABITS: Red hake feed primarily on benthic crustaceans, particularly small crabs and shrimps. Fishes, squids, bivalves, and worms also appear in their diet.

IMPORTANCE: Red hake are of no commercial or recreational importance in the Chesapeake Bay region.

Spotted hake - *Urophycis regia* (Walbaum, 1792)

KEY FEATURES: First dorsal fin small, without trailing filament; pelvic fin modified into forked pair of filaments, the longer of which reaches anal fin. **COLOR:** Back and sides pale brown, white on belly; lateral line dark, punctuated with series of white spots; first dorsal fin pale at base, black above, and margined with white; black spots on head; pelvic fins white. **SIZE:** Maximum adult size about 41 cm (1.4 ft) TL. **RANGE:** Southern New England to Florida and the northern Gulf of Mexico. Juveniles occasionally found as far north as Nova Scotia.

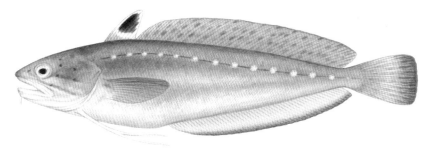

Spotted hake *Urophycis regia*

HABITAT AND HABITS: Spotted hake are a warm-temperate coastal species that migrates inshore and north in the spring in the Mid-Atlantic Bight and offshore and south in the fall, overwintering primarily south of the Chesapeake Bay. They are benthic fish that usually occur over soft substrates and are typically associated with objects on the bottom. **OCCURRENCE IN THE CHESAPEAKE BAY:** Juvenile spotted hake are common in the lower Chesapeake Bay and its tributaries and occasional in the upper bay from March to June, leaving the bay as water temperatures approach 25°C (77°F). They may move up tributaries to waters of 8–10‰ and rarely almost to freshwater.

REPRODUCTION: Spotted hake probably mature at age 1 or 2, based on size at maturity (21–31 cm, or 8–12 in). Maximum age is unknown. Spawning occurs on the continental shelf and upper slope at two primary periods in our area: August to November and February to April. The eggs and larvae are pelagic.

FOOD HABITS: Juvenile spotted hake feed on small crustaceans such as copepods and amphipods. Larger spotted hake consume decapods, fishes, squids, and worms.

IMPORTANCE: Spotted hake are too small to have any value to recreational or commercial fisheries.

Oyster toadfish - *Opsanus tau* (Linnaeus, 1766)

KEY FEATURES: Fleshy flaps on cheeks and jaws; mouth very broad, with strong blunt teeth; gill cover with two partly concealed spines; pectoral fins broad, axil of pectoral fin with large blind pouch. **COLOR:** Body yellowish to olive brown with darker blotches; may be darker or lighter depending on bottom color; pectoral fin with dark concentric bands. **SIZE:** Maximum adult size 38 cm (1.3 ft) TL. **RANGE:** Cape Cod (straying as far north as Maine) to Florida.

Oyster toadfish *Opsanus tau*

HABITAT AND HABITS: Oyster toadfish are benthic residents in estuaries and occur over a wide range of temperatures, salinities, depths, and bottom types from spring through autumn. Winter records are sparse, and it is assumed that oyster toadfish bury themselves in mud in deeper portions of estuaries at that time. **OCCURRENCE IN THE CHESAPEAKE BAY:** Oyster toadfish are abundant and ubiquitous in the Chesapeake Bay but are most common in the middle and lower portion of the bay and in tributaries on mud, rock, sand, and oyster-shell bottoms.

REPRODUCTION: Age at maturity of toadfish has not been recorded, but it is probably two or three years. Maximum age is 9 to 12 years. Male oyster toadfish occupy nest sites under large shells, logs, or other debris or in old tin cans from late spring through mid-summer. There they emit grunts and a typical "boatwhistle" call that serves to attract females and to discourage other males from intruding into their territory. The males spawn with one or more females and then remain at the nest to guard the sticky benthic eggs and young, even after they have been hatched for a week or two.

FOOD HABITS: Oyster toadfish are sluggish demersal predators that feed primarily on small crabs and shrimps.

IMPORTANCE: There are no commercial or recreational fisheries for "oyster toads." They are often taken inadvertently by recreational fishers while bottom-fishing with peeler crab or cut bait for more desirable species. Care must be taken when unhooking oyster toadfish, as they are quick to bite unwary fingers, and the pungent fin and opercular spines can cause painful punctures.

Goosefish - *Lophius americanus* Valenciennes, 1837

KEY FEATURES: Large flat head with huge mouth, projecting lower jaw filled with large sharp teeth; fleshy flaps around head and body; top of head with three long modified fin spines, the first of which, called an illicium, is tipped with a small irregular fleshy appendage (esca) that acts as a fishing lure. **COLOR:** Chocolate brown with some pale and dark mottling above; color shade depends on substrate color; dirty white below. **SIZE:** Maximum adult size 1.2 m (4 ft) TL. **RANGE:** Grand Banks of Newfoundland to North Carolina, south to Florida in deep water.

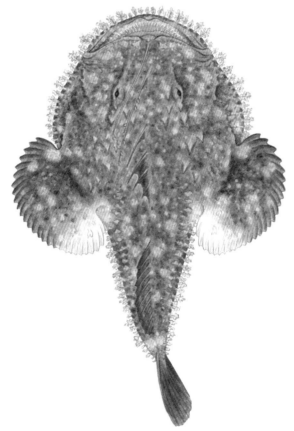

Goosefish *Lophius americanus*

HABITAT AND HABITS: Goosefish are marine, bottom-dwelling fishes. Goosefish inhabit sand, mud, and broken-shell bottoms from a few centimeters in depth to over

800 m (2,616 ft). Goosefish are cold-water coastal fish that undertake seasonal migrations, south and inshore in the colder months and north and/or offshore in the warmer months. OCCURRENCE IN THE CHESAPEAKE BAY: Goosefish are occasional visitors to the lower Chesapeake Bay from late autumn to early spring. They are most abundant in waters with temperatures between 3 and 11°C (37–52°F); thus they occur in the lower bay only during the colder months.

REPRODUCTION: Goosefish mature at age 3 or 4 and may live for 9 to 11 years. Spawning occurs on the continental shelf in spring. Females produce long veil-like strings of eggs in a floating grayish gelatinous matrix. These veils may contain more than 1.3 million eggs. The principal nursery area for goosefish in the Mid-Atlantic Bight appears to be on the upper continental slope.

FOOD HABITS: Aided by their fishing appendage, goosefish are voracious lie-in-wait benthic predators. Goosefish sit on the bottom, wiggling their esca to attract prey close to their large mouth, where the prey is engulfed. They are capable of eating prey fully half as large as they are because of their capacious mouth and stomach. Goosefish prey on fishes, squids, and various kinds of sea birds, most of which they must take on the surface.

IMPORTANCE: Goosefish meat is firm, white, and sweet and is marketed as monkfish. Goosefish are occasionally taken by recreational fishers on artificial lures while fishing for other species, such as striped bass or bluefish.

Striped mullet - *Mugil cephalus* Linnaeus, 1758

KEY FEATURES: Anal fin with three spines and eight rays; second dorsal and anal fins with few or no scales. **COLOR:** Dark gray to blue on back, silvery white on sides and belly; dark spot on individual scales along sides, forming lateral stripes; dark blotch at base of pectoral fin; small juveniles silvery. **SIZE:** Maximum adult size 91 cm (3 ft) TL but rarely larger than 55 cm (1.8 ft) TL. **RANGE:** Worldwide in warm-temperate to tropical latitudes. In the western North Atlantic from Cape Cod (summer) to Brazil, with juveniles straying to Nova Scotia.

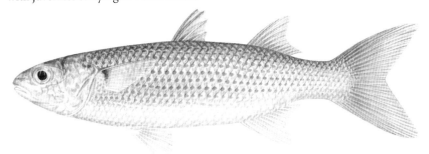

Striped mullet *Mugil cephalus*

HABITAT AND HABITS: Striped mullet are coastal fishes that readily penetrate estuaries and coastal rivers into freshwater. Pelagic larvae and juveniles are carried up into Mid-Atlantic Bight waters by the Gulf Stream and recruit into estuaries as far north as Cape Cod in summer. In autumn they form large schools and migrate down the coast to south of Cape Hatteras. Older juveniles and adults regularly migrate north into mid-Atlantic estuaries in summer and return south in autumn to winter offshore. Mullet are known for their habit of frequently jumping well clear of the water, a habit that has resulted in the frequently used name "jumper." **OCCURRENCE IN THE CHESAPEAKE BAY:** Striped mullet are abundant visitors from spring through autumn throughout the Chesapeake Bay and occur in a wide variety of habitats from the main stem, ascending up the major tributaries to freshwater and into small tidal marsh creeks.

REPRODUCTION: Striped mullet mature at age 3 and may live for 16 years. Spawning occurs in large schools in winter, well offshore, somewhere in the South Atlantic Bight. Females may produce half a million to two million eggs.

FOOD HABITS: Striped mullet usually feed on the bottom on detritus and microscopic algae. They have a muscular gizzard-like stomach and long coiled intestine particularly well suited for digesting plant material.

IMPORTANCE: Striped mullet are of minor importance to fisheries in the Chesapeake Bay but are very important elsewhere. Mullet are eaten in the South and play a major role as bait in a multitude of recreational fisheries from the Middle Atlantic coast to Texas. Adult mullet are used for cut bait and smaller mullet, called "finger mullet," may be used whole.

White mullet - *Mugil curema* Valenciennes, 1836

KEY FEATURES: Anal fin with three spines and nine rays; second dorsal and anal fins densely covered with fine scales, nearly to fin margin. **COLOR:** Dark gray to blue on back, silvery white on sides and belly; no stripes on sides; dark blotch at base of pectoral fin; gill cover with distinct yellow-gold spot; small juveniles silvery. **SIZE:** Maximum adult size 45 cm (1.5 ft) TL, but typically less than 35 cm (1.2 ft) TL. **RANGE:** Eastern and western Atlantic and eastern Pacific. In the western North Atlantic from Cape Cod (summer) to Brazil, with juveniles straying to Nova Scotia.

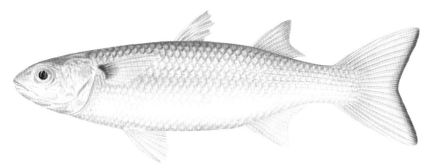

White mullet *Mugil curema*

HABITAT AND HABITS: White mullet are coastal fishes that enter estuaries and prefer higher salinity areas (>8‰). Pelagic larvae and juveniles carried up into Mid-Atlantic Bight waters by the Gulf Stream recruit into estuaries as far north as Cape Cod in summer. In autumn they migrate down the coast to south of Cape Hatteras. Older juveniles and adults are rarely found north of Florida. **OCCURRENCE IN THE CHESAPEAKE BAY:** Juvenile white mullet are common summer and autumn visitors to the lower bay and are occasionally found in the middle bay.

REPRODUCTION: White mullet mature at about age two and may live for at least five years. Spawning occurs in large schools in spring, well offshore, somewhere in the South Atlantic Bight. Eggs and larvae are pelagic.

FOOD HABITS: White mullet usually feed on the bottom on detritus and microscopic algae and have a muscular gizzard-like stomach and a long coiled intestine particularly well-suited for digesting plant material.

IMPORTANCE: Due to their more southerly distribution and smaller size, white mullet have little fisheries importance along the Atlantic seaboard, with the exception of Florida, where modest numbers are landed.

Rough silverside - *Membras martinica*
(Valenciennes, 1835)

KEY FEATURES: Scales firm, fringed along rear margin, rough to touch; base of dorsal and anal fins covered with large scales. COLOR: Translucent blue to green on back, silvery on sides and belly, with brilliant silver stripe along side. SIZE: Maximum adult size 12.5 cm (4.9 in) SL, commonly exceeding 7.5 cm (3 in) SL. RANGE: Inhabiting coastal areas from New York to Yucatán, Mexico.

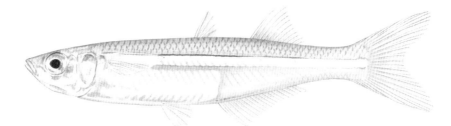

Rough silverside *Membras martinica*

HABITAT AND HABITS: Rough silversides are a pelagic schooling species that occurs in brackish to highly saline open-water habitats along beaches and channel edges, from spring to autumn. In winter, they move into deeper estuarine or nearby marine waters. This species typically schools near the surface, from shallow flats to water 3–15 m (10–49 ft) deep. OCCURRENCE IN THE CHESAPEAKE BAY: Rough silversides occur from spring to autumn, in mid- to high-salinity areas of the Chesapeake Bay, but have been taken in tidal rivers at salinities as low as 8‰. Unlike other silverside species in our area, rough silversides seem to avoid marsh habitats and are more often found along exposed shorelines and beaches or in deeper areas in the bay or its tributaries. They may be the most common of our silversides taken by dipnet under dock lights at night. In winter, rough silversides move into deep higher-salinity waters of the lower bay and nearby ocean.

REPRODUCTION: Rough silversides mature in a year and probably live for two or perhaps three years. They spawn sticky demersal eggs over submerged grass beds in summer.

FOOD HABITS: Rough silversides feed on copepods and other planktonic crustaceans.

IMPORTANCE: Rough silversides have no importance in commercial fisheries in the Chesapeake Bay, although they may be used for bait by sport fishers. However, these common pelagic schooling species are important components of bay ecosystems, where they are preyed on by several larger fish species, such as striped bass.

Inland silverside - *Menidia beryllina* (Cope, 1867)

KEY FEATURES: Scales smooth, not fringed along rear margin; 36–42 scale rows; first dorsal-fin origin in front of vent; 13–19 (typically 16) anal-fin soft rays; base of dorsal and anal fins without scales. COLOR: Greenish on back, silvery white on side and belly; distinct silver stripe on side. SIZE: Maximum adult size 7.5 cm (3 in) TL. RANGE: Massachusetts to Vera Cruz, Mexico.

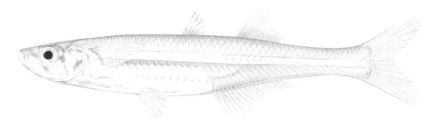

Inland silverside *Menidia beryllina*

HABITAT AND HABITS: Inland silversides are a schooling species found in freshwater and brackish marsh creeks and pools during the warmer months. In winter they move into deeper, more saline parts of estuaries. OCCURRENCE IN THE CHESAPEAKE BAY: Inland silversides are common to abundant in freshwater and brackish marshes and grass beds throughout the bay from spring through autumn. They usually are found at lower salinities than the other two members of the family that occur in the bay. In winter, they move into deep water in the lower bay.

REPRODUCTION: Inland silversides mature in one year, and probably few survive to age two. Inland silversides spawn in spring and summer in fresh and brackish water. They deposit their eggs on vegetation, detritus, roots, or leaves in the intertidal zone at high tide during the day. Females produce multiple batches of eggs over a protracted spawning period and may spawn on a daily basis, producing as many as 150,000 eggs during the season.

FOOD HABITS: Inland silversides are daytime predators that form dense schools and primarily locate prey by sight. Inland silversides feed on small crustaceans such as copepods and shrimps.

IMPORTANCE: Inland silversides are an important forage species but have no commercial or recreational value except as occasional bait for sport fishers.

Atlantic silverside - *Menidia menidia* (Linnaeus, 1766)

KEY FEATURES: Scales smooth, not fringed along rear margin; 43–55 scale rows; first dorsal-fin origin behind vent; 19–29 (typically 23–25) anal-fin soft rays; base of dorsal and anal fins without scales. COLOR: Greenish on back, silvery white on side and belly; distinct silver stripe on side. SIZE: Maximum adult size 12.7 cm (5 in) TL. RANGE: Inhabiting coastal areas from Nova Scotia to Florida.

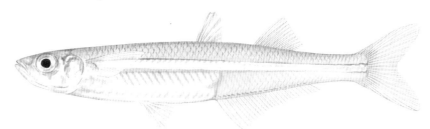

Atlantic silverside *Menidia menidia*

HABITAT AND HABITS: Atlantic silversides are a schooling species found typically in mid- to high-salinity areas over grass beds, along sandy beaches, and in marsh creeks during the warmer months. In winter, they move into deeper nearshore ocean waters. Atlantic silversides usually travel in schools of a few dozen to several hundred fish. OCCURRENCE IN THE CHESAPEAKE BAY: Atlantic silversides are abundant throughout the Chesapeake Bay, usually close to shore in the habitats mentioned above. Although they occur in greatest abundance at higher salinities, they occasionally penetrate sufficiently far upstream in the tributaries to enter freshwater. In winter, large schools of Atlantic silversides move into deep, more saline water at the bay mouth and in the coastal ocean.

REPRODUCTION: Atlantic silversides mature in one year, and few reach age two. This silverside is one of many species known to have lunar-related spawning cycles. Spawning occurs strictly during daylight hours in large schools and coincides with high tide. The first spawning activity usually occurs at a new or full moon, followed by spawning peaks at two-week intervals. Spawning occurs from March to July in the intertidal zone or shallow estuarine waters. The eggs are attached to marsh grass or filamentous algae and are adapted to be exposed to air for several hours a day.

FOOD HABITS: Atlantic silversides mostly consume small crustacean zooplankters, such as copepods and shrimps, but also invertebrate and fish eggs and larvae and diatoms (whether they can digest diatoms is unknown). When near shore, these silversides feed high in the water column on copepods, whereas in winter when in deeper water, they feed closer to the bottom, primarily on shrimps. Opportunistic omnivores, they cluster in feeding schools, following the tidal ebb and flow along shorelines.

IMPORTANCE: Atlantic silversides are a particularly important forage species because of their abundance and schooling habit. A wide variety of piscivorous fishes and birds feed heavily on silversides. Atlantic silversides are a popular bait in the summer flounder recreational fishery and are usually sold frozen, several to a container, and called "spearing."

Flat needlefish - *Ablennes hians* (Valenciennes, 1846)

KEY FEATURES: Body long and greatly compressed; rear of dorsal fin extended into distinct lobe; 24–28 (usually 26–27) anal-fin rays; pectoral fin pointed, sickle-shaped. COLOR: Dark bluish green on back, silvery on sides and belly; dark bars along side; lower jaw with red tip; dorsal-fin rays tipped with black, rear dorsal-fin lobe

Flat needlefish *Ablennes hians*

black. SIZE: Maximum adult size 83 cm (2.7 ft) SL. RANGE: Worldwide in warm offshore waters. In the western Atlantic, the Chesapeake Bay (straying as far north as Cape Cod) to Brazil.

HABITAT AND HABITS: Flat needlefish are offshore pelagic surface-living fish that occur primarily in subtropical to tropical regions but extend their range to warm-temperate latitudes in summer. They appear to be most common around tropical islands but may enter estuaries and coastal rivers. OCCURRENCE IN THE CHESAPEAKE BAY: Flat needlefish are occasional summer visitors to the lower Chesapeake Bay, reaching as far north as the Potomac River. Most specimens encountered are juveniles. They may occasionally penetrate up the larger tributaries to salinities as low as 18‰.

REPRODUCTION: Flat needlefish spawn offshore, probably in summer. A single female was reported to have more than 600 eggs with adhesive tufts in her ovary. This suggests that the eggs are attached to vegetation, as with other species in the family. Fecundity may be higher than suggested by the ovarian content because more than one batch of eggs might be produced during the spawning season.

FOOD HABITS: Flat needlefish feed on small pelagic fishes.

IMPORTANCE: There is no commercial interest in flat needlefish in the United States, although recreational fishers occasionally catch them while trolling for other species.

Atlantic needlefish - *Strongylura marina*
(Walbaum, 1792)

KEY FEATURES: Body oval in cross-section; 17 or fewer dorsal-fin rays; tail weakly forked, no keel in front of tail; pectoral fin rounded, not sickle-shaped. COLOR: Back green, silvery along side, lateral blue stripe, belly white. SIZE: Maximum adult size 64 cm (2.1 ft) SL. RANGE: Massachusetts to the Gulf of Mexico, Central America to Brazil. Absent from the Bahamas and Antilles.

Atlantic needlefish *Strongylura marina*

HABITAT AND HABITS: Atlantic needlefish occur in coastal areas from warm-temperate to tropical latitudes. They enter estuaries, where they may move almost to freshwater and typically occur close to shore over grass beds and along beaches. Atlantic needlefish are conspicuous fish, due to their habit of swimming near the surface and being attracted to lighted piers and bridges. OCCURRENCE IN THE CHESAPEAKE BAY: Atlantic needlefish are common from spring to autumn throughout the Chesapeake Bay, reaching as far north as the Susquehanna River. They typically ascend tributaries as far as freshwater. Young Atlantic needlefish as small as 5 cm (2 in) are frequently found in summer in grass beds and along marsh margins from the bay mouth to the Patuxent River.

REPRODUCTION: Based on growth patterns and size at maturity, Atlantic needlefish probably mature at age one and live for two or three years. Spawning occurs on seaweeds or grass beds in spring and early summer, in estuaries and river mouths as far upstream as freshwater. A single female was recorded as containing 1,000 eggs in her ovary, but annual fecundity may be higher, because more than one batch of eggs might be produced per season.

FOOD HABITS: Atlantic needlefish prey on small fishes such as silversides, which they catch sideways in their jaws and then turn to swallow (much like fish-eating birds).

IMPORTANCE: Atlantic needlefish have no commercial or recreational value in the Chesapeake Bay region.

Houndfish - *Tylosurus crocodilus* (Péron and Lesueur, 1821)

KEY FEATURES: Body oval in cross-section; 21–23 dorsal-fin rays; tail strongly forked, lower lobe much longer than upper; small dark keel on side in front of tail; 18–22 anal-fin rays. COLOR: Back dark bluish green, silvery along side, lateral blue stripe, belly white; scales and bones greenish. SIZE: Maximum adult size 1.3 m (4.3 ft) TL. RANGE: Worldwide in warm coastal waters. In the western Atlantic, the Chesapeake Bay to Brazil.

Houndfish *Tylosurus crocodilus*

HABITAT AND HABITS: Houndfish occur in subtropical and tropical coastal waters, extending into warm-temperate areas in summer. **OCCURRENCE IN THE CHESA-PEAKE BAY:** Houndfish occur throughout the Chesapeake Bay in summer. They are particularly common over grass flats in the lower bay but extend as far north in the bay as the lower Susquehanna River basin.

REPRODUCTION: Little is known about reproduction in houndfish, other than that one large female from the Virgin Islands was recorded as containing 25,000 ripe eggs. As with other members of this family, the eggs have adhesive filaments and are assumed to be deposited on vegetation. Spawning probably occurs principally in subtropical and tropical areas.

FOOD HABITS: Houndfish are active surface piscivores.

IMPORTANCE: Houndfish will readily strike a rapidly stripped streamer fly but are difficult to hook because of their hard mouth. A wisp of nylon stocking threaded on the fly's hook will improve the chance of hookup, as the fish become entangled by their needle-like teeth. Once hooked, houndfish are spectacular aerial fighters. Their proclivity to jump when startled has occasionally led to injuries to boaters who are impaled while speeding over the flats.

Ballyhoo - *Hemiramphus brasiliensis* (Linnaeus, 1758)

KEY FEATURES: Origin of dorsal fin in front of anal fin; tail deeply forked, lower lobe longer than upper; pelvic fin much closer to tail than to gill opening. **COLOR:** Dark blue to green on back, silvery on sides, and white below; upper lobe of tail yellow orange, lower lobe dusky; beak black with fleshy red tip. **SIZE:** Maximum adult size 55 cm (1.8 ft) TL, but typically 35 cm (1.1 ft) TL or less. **RANGE:** Ballyhoos occur in both the eastern and western Atlantic. In the latter, they range from Massachusetts to Brazil, including the Gulf of Mexico and Caribbean.

Ballyhoo *Hemiramphus brasiliensis*

HABITAT AND HABITS: Ballyhoos are a coastal pelagic surface-schooling species, most abundant in subtropical and tropical regions, but entering warm-temperate waters in summer. OCCURRENCE IN THE CHESAPEAKE BAY: Ballyhoos are rare summertime visitors to the lower Chesapeake Bay.

REPRODUCTION: Ballyhoos mature in one year and rarely exceed two years of age. Their reproductive habits are little known, but other members of the genus have eggs with adhesive tendrils that attach to sea grasses or floating vegetation.

FOOD HABITS: Ballyhoos are mostly vegetarian and feed on bits of floating plant matter. They also have been reported to take small herrings.

IMPORTANCE: Ballyhoos are taken by dip-netting under lights at night, by cast nets, and by drag nets.

False silverstripe halfbeak - *Hyporhamphus meeki*
Banford and Collette, 1993

KEY FEATURES: Dorsal-fin origin on a vertical with, or slightly in front of, anal-fin origin; tail forked, lower lobe slightly longer than upper; pelvic fin short, about midway between tail and gill opening. COLOR: Dark blue to green on back, silvery on sides, and white below; three dark stripes on back between head and dorsal fin; tail pale, with dark edge; red fleshy tip on beak. SIZE: Maximum adult size 22 cm (8.6 in) TL, but commonly to 18 cm (7 in) TL. RANGE: These halfbeaks occur from Cape Cod to Florida and the Gulf of Mexico to Yucatán.

False silverstripe halfbeak *Hyporhamphus meeki*

HABITAT AND HABITS: False silverstripe halfbeaks are a coastal pelagic surface-schooling species, most abundant in warm-temperate regions in summer and in subtropical regions year-round. They commonly enter estuaries and are typically found over sandy vegetated bottoms. OCCURRENCE IN THE CHESAPEAKE BAY: These halfbeaks are common summer and autumn visitors to the Chesapeake Bay, extending as far north as the Chester River. This species occurs in small schools over eelgrass beds and along beaches. They are attracted to lights at night but are quick and agile swimmers that are difficult to capture with a dip-net.

REPRODUCTION: Not much is known about the age of maturity in this species, but it almost surely is similar to ballyhoos, which mature at age one and rarely exceed age two. Spawning occurs in the summer, and the adhesive eggs are attached to floating sea grasses.

FOOD HABITS: False silverstripe halfbeaks feed on bits of floating vegetation and detritus and small invertebrates.

IMPORTANCE: These halfbeaks have little commercial or recreational value because of their small size (about half that of the widely-used baitfish ballyhoos). They can be collected by seines and dip-nets.

Sheepshead minnow - *Cyprinodon variegatus*
Lacepède, 1803

KEY FEATURES: Short chubby body; snout short, about equal to eye diameter; teeth incisor-like, in single row; large scale above pectoral fin at rear of gill opening; 24–27 scale rows. COLOR: Females are brassy above and on sides, with dark blotches forming bars low on side; rear of dorsal fin with "eye spot"; males are olive above, suffused with iridescent blue specks, bright orange below during breeding season; dorsal fin dark, with orange margin; tail greenish, with dark bar at base. SIZE: Males are larger than females; maximum adult size 7.6 cm (3 in) TL. RANGE: Fresh and brackish waters from Cape Cod to Mexico.

HABITAT AND HABITS: Sheepshead minnows occupy shallow estuarine habitats and often travel in large schools. They are a hardy species that has been found from freshwater to salinities exceeding 90‰. Sheepshead minnows are particularly abundant in salt marsh creeks and ponds and may forage on the marsh surface at high tide. In winter, sheepshead minnows may move into deeper water or remain buried in silt in marsh creeks and ponds. OCCURRENCE IN THE CHESAPEAKE BAY: Sheepshead minnows are abundant throughout the bay, where they frequent shallow flats, marshes, and tidal ponds during the summer months and retreat to channels or burrow into the silt in marsh ponds in the winter. They are often found over a substrate of thick mud and detritus.

female

male

Sheepshead minnow *Cyprinodon variegatus*

REPRODUCTION: Sheepshead minnows mature at age one and live for two or occasionally three years. Spawning occurs in spring and summer, primarily in vegetated marsh pools. The eggs are demersal and have sticky filaments.

FOOD HABITS: Sheepshead minnows consume detritus and small crustaceans.

IMPORTANCE: Sheepshead minnows are of no recreational or commercial interest but are an important forage species for other fishes and wading birds, particularly in marsh habitats.

Banded killifish - *Fundulus diaphanus* (Lesueur, 1817)

KEY FEATURES: Teeth in multiple rows; 35–52 scale rows, but usually more than 40. **COLOR:** Olive on back, silvery white on sides, white on belly; small dark spot on gill cover behind eye, 12–20 dark bars on sides, bars wider on males, about the same width as the interspaces; female bars narrow and shorter than those on males. **SIZE:** Maximum adult size 11 cm (4.3 in) TL, but typically less than 8.5 cm (3.3 in) TL. **RANGE:** Banded killifish are found in the upper Mississippi and Great Lakes drainages and on the Atlantic slope from Newfoundland to South Carolina.

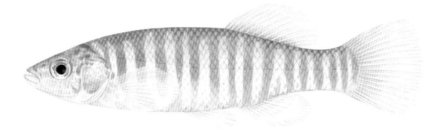

Banded killifish *Fundulus diaphanus*

HABITAT AND HABITS: Banded killifish occur in a wide variety of shallow freshwater and brackish habitats, from ponds and streams to marshes, beaches, and rivers. This species tends to avoid acidic waters. **OCCURRENCE IN THE CHESAPEAKE BAY:** Banded killifish are common to abundant in tributaries throughout the bay, primarily in fresh and slightly brackish water, but rarely to salinities as high as 19‰. They are more likely to be found in freshwaters than other members of the genus. While banded killifish are most abundant on the Coastal Plain, they are also found on the Piedmont, and montane populations are known in at least two river systems.

REPRODUCTION: Banded killifish mature at age one and live to age three. Spawning occurs from April to September, and females deposit about 10 adhesive eggs per batch, usually in aquatic vegetation. Total fecundity has been estimated at 200–250 eggs.

FOOD HABITS: Banded killifish feed on small crustaceans, mollusks, and worms.

IMPORTANCE: Banded killifish have no commercial value but are sometimes used as bait by recreational fishers. They are of considerable value indirectly as a food source for other fishes and wading birds.

Mummichog - *Fundulus heteroclitus* (Linnaeus, 1766)

KEY FEATURES: Snout about equal to eye diameter; teeth in multiple rows; gill opening not restricted, extending to upper rear of gill cover; typically 31–39 scale rows; dorsal-fin base longer than anal-fin base; 10–15 dorsal-fin rays; 9–12 anal-fin rays. **COLOR:** Females dark green on back, lighter on side, with 12–15 dusky bars, belly pale; males olive on back, silvery on side, with 15 light bars and numerous pale spots, belly yellowish; dorsal fin often with black spot at rear; during breeding season males become brighter, with bright blue cast to back and pearly bars and spots on side. **SIZE:** Maximum adult size 12 cm (4.7 in) TL, with females attaining larger sizes than males. **RANGE:** Coastal waters and rivers from Newfoundland to northern Florida.

HABITAT AND HABITS: Mummichogs occupy a wide variety of shallow estuarine habitats, from seagrass beds to marsh creeks, foraging on the marsh surface at high tide. They are among the hardiest of fishes, occupying a wide range of salinities, from freshwater to hypersaline marsh pools, and temperatures greater than 34°C (93°F) to 0°C (32°F). In winter, they have been reported to move into deeper water but also to burrow into subtidal silt and to form swarms under the ice in small tributaries close to tidal water. Mummichogs prefer lower salinities than do striped killifish and are much more likely to be found in tidal freshwater. They are usually encountered in schools of a few to several hundred individuals. **OCCURRENCE IN THE CHESAPEAKE**

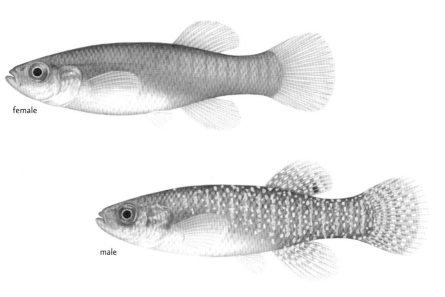

female

male

Mummichog *Fundulus heteroclitus*

BAY: Mummichogs are ubiquitous and one of the most abundant fishes in the Chesapeake Bay.

REPRODUCTION: Mummichogs mature at age one and live for three years. Spawning occurs from April to August during spring tides, on the marsh surface. Eggs are resistant to desiccation and are usually deposited in empty mussel shells or similar sheltered places. Fecundity has been reported to be 460–800 mature eggs, usually deposited over a 3- or 4-day period around the spring tides. However, females may develop additional batches of eggs between spring tides over the course of the long spawning season, so that total fecundity might be much higher.

FOOD HABITS: Mummichogs consume small crustaceans, snails, worms, insects, and other invertebrates and occasionally take small fishes.

IMPORTANCE: Mummichogs are widely used as bait in recreational fisheries, both in salt and freshwater, and are taken in minnow pots and by beach seines. They are among the most important forage fishes in the Chesapeake Bay.

Spotfin killifish - *Fundulus luciae* (Baird, 1855)

KEY FEATURES: Snout short, about equal to or less than eye diameter; teeth in multiple rows; gill opening restricted, extending only to upper part of pectoral-fin base; 34–36 scale rows; dorsal-fin base shorter than anal-fin base; 8–9 dorsal-fin rays. **COLOR:** Both sexes greenish gray on back, with lighter side and belly; males brighter during breeding season, with golden and yellow overtones on side and belly, 11–14 dark bars on side, and black ocellus on rear of dorsal fin. **SIZE:** These are the smallest species of killifish, with a maximum adult size of 5 cm (2 in) TL. **RANGE:** Intertidal marshes from Long Island and Connecticut to Georgia.

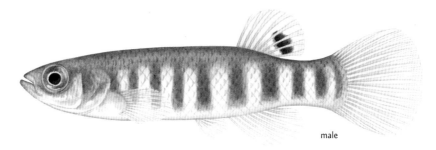

male

Spotfin killifish *Fundulus luciae*

HABITAT AND HABITS: Spotfin killifish are rarely encountered, probably due to their small size and preferred habitat; they are restricted to shallow intertidal high-marsh habitats, where they may be subjected to a wide range of salinities and temperatures. They are elusive, living among the interstices of marsh grasses and crab burrows and in small marsh pools. **OCCURRENCE IN THE CHESAPEAKE BAY:** Spotfin killifish are locally abundant permanent residents of the Chesapeake Bay that are found in tidal rivulets and puddles in the upper reaches of intertidal marshes. They probably overwinter by burrowing into silty marsh sediments.

REPRODUCTION: Spotfin killifish mature at age one, and few survive to age two. Spawning occurs from April to August. As with other killifishes, the eggs are adhesive, benthic, resistant to desiccation in air, and probably deposited on the marsh surface.

FOOD HABITS: Marsh killifish are omnivorous and feed on detritus, diatoms, and a variety of small crustaceans, insects, and worms.

IMPORTANCE: Spotfin killifish are of no commercial or recreational interest. As a primary consumer, they are an important link in the high-marsh food web.

Striped killifish - *Fundulus majalis* (Walbaum, 1792)

KEY FEATURES: Snout long, about twice eye diameter; teeth in multiple rows; gill opening not restricted, extending to upper rear of gill cover; typically 31–38 scale rows; dorsal-fin base longer than anal-fin base; 11–16 dorsal-fin rays. **COLOR:** Females olive above and white below, with 2 or 3 irregular black stripes on side and small number of black vertical bars near tail; males olive above and yellowish on side and belly, with 15–20 black vertical bars on sides. **SIZE:** Maximum adult size 20 cm (8 in) TL, with females attaining larger sizes than males. These are the largest species of killifish in the bay. **RANGE:** Coastal waters and river mouths from New Hampshire to northeast Florida and the northern Gulf of Mexico.

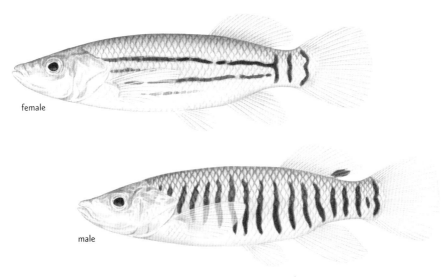

female

male

Striped killifish *Fundulus majalis*

HABITAT AND HABITS: Striped killifish prefer higher salinities (meso-polyhaline conditions) and rarely, if ever, enter freshwater. They are found in tidal creeks, sand flats, and grass beds. Striped killifish are often seen in large schools swimming in shallow waters, sometimes in depths of only a few inches. They are reported to burrow in mud during the coldest months. **OCCURRENCE IN THE CHESAPEAKE BAY:** Striped killifish are abundant permanent residents of the entire bay. They tend to occur over sandy sediments more often than do other local killifishes.

REPRODUCTION: Striped killifish mature at age two and have been reported to live for six or seven years. However, these ages have been disputed, and a usual maximum age of four years may be more realistic. Spawning occurs in quiet areas from April to September, the females actively burying their eggs in the sand. Fecundity has been reported at 200–800 eggs, but the species is a known batch spawner, with several spawning peaks during the season, so total annual fecundity may be higher.

FOOD HABITS: Striped killifish consume worms, small crustaceans, mollusks, and insects.

IMPORTANCE: Striped killifish are sometimes used as bait by recreational fishers and are taken in minnow traps, seines, and cast nets. They are important forage fish in higher-salinity areas in the bay.

Bayou killifish - *Fundulus pulvereus* (Evermann, 1892)

KEY FEATURES: Snout short, about equal to eye diameter; teeth in multiple rows; gill opening not restricted, extending to upper rear of gill cover; 37 or fewer scale rows; dorsal-fin base about equal to or only slightly longer than anal-fin base; 10 dorsal-fin rays; 9–11 anal-fin rays. COLOR: Females, olive brown on back, gold on sides, and silvery white on belly; sides with numerous black spots, sometimes merging to form horizontal lines; males, olive to light brown on back, silvery on side and belly, with 12–17 vertical bars on side, about same width as interspaces; sides behind dorsal and anal fins with pearly light spots. SIZE: Maximum adult size 6.5 cm (2.6 in) TL.

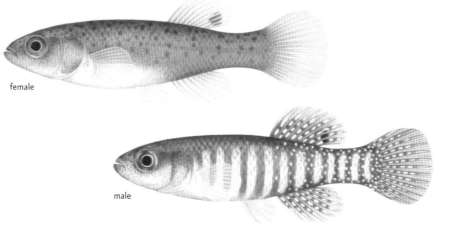

female

male

Bayou killifish *Fundulus pulvereus*

RANGE: Bayou killifish are found from the lower Chesapeake Bay to the St. Johns River in northern Florida and along the Gulf coast from Alabama to Corpus Christi, Texas. This species is replaced on the Florida peninsula by the closely related marsh killifish, *Fundulus confluentus*, with which it was long confused. Earlier records of the marsh killifish from the Chesapeake Bay appear to be based on the bayou killifish.

HABITAT AND HABITS: Bayou killifish are found in grassy backwaters, brackish bayous, tidal creeks and bays, barrier beach island ponds, and freshwater. They have a very wide salinity tolerance. OCCURRENCE IN THE CHESAPEAKE BAY: Bayou killifish are rarely encountered in the Chesapeake Bay and are reported only as far north as the mouth of the York River. They have been well documented only in Lynnhaven Bay near Cape Henry, where they appear to have been extirpated due to habitat changes.

REPRODUCTION: Bayou killifish mature at age one and may live for two or three years. Spawning can occur from spring to autumn, usually during biweekly lunar high tides. Females deposit about 40 adhesive eggs per batch at the base of vegetation or in crevices of various kinds of debris in the intertidal zone. The eggs are extremely resistant to aerial desiccation.

FOOD HABITS: Bayou killifish feed on insect larvae, including mosquito larvae, small grass shrimps, and snails.

IMPORTANCE: Bayou killifish are of no importance to recreational or commercial fisheries but may play a role in controlling mosquitoes.

Rainwater killifish - *Lucania parva*
(Baird and Girard, 1855)

KEY FEATURES: Teeth pointed, in single row; 25–26 scale rows. COLOR: Olive above, shading to pearly white below. During breeding season males may have orange to reddish tinge to fins; dorsal fin sometimes with black anterior spot. SIZE: Maximum adult size 6.2 cm (2.4 in) TL. RANGE: Brackish and coastal waters from Cape Cod to the Florida Keys and the Gulf of Mexico to Veracruz.

HABITAT AND HABITS: Rainwater killifish occupy a wide salinity range, from freshwater to full seawater, but are restricted mostly to shallow grass beds, where they are usually found in small schools. Rainwater killifish may exhibit cleaning symbiosis with others of the same species or with sheepshead minnows. This complex

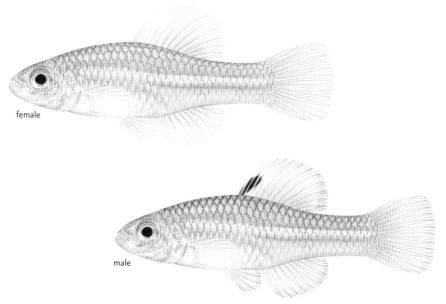

female

male

Rainwater killifish *Lucania parva*

behavior results in one fish picking the parasites from the skin of another fish. OC-CURRENCE IN THE CHESAPEAKE BAY: Rainwater killifish are ubiquitous in shallow grass beds throughout the bay at least as far north as the Chester River. They are year-round residents. During the summer they are abundant in all brackish water habitats where vegetation is present. During the winter, they burrow into the bottom silt in low-salinity tidal ponds. This species travels in schools and is often found in association with other topminnows and killifishes.

REPRODUCTION: Rainwater killifish mature at age one and rarely attain two years of age. Spawning occurs from April to July, the adhesive demersal eggs being deposited in shallow water in seagrasses or on algal mats, where males set up and defend territories. Individual females deposit several batches of eggs during the season.

FOOD HABITS: Rainwater killifish feed mostly on larval crustaceans, small worms, mollusks, mosquito larvae, and grass shrimps.

IMPORTANCE: Rainwater killifish have no recreational or commercial value but are an important forage species in grass beds.

Eastern mosquitofish - *Gambusia holbrooki*
Girard, 1859

KEY FEATURES: Single dorsal fin placed far back on body, dorsal-fin origin well behind anal-fin origin; dorsal-fin base shorter than anal-fin base in females; anal fin similar to dorsal fin in females, in males third to fifth anal-fin rays modified as intromittent organ (gonopodium). **COLOR:** Olive on back, grayish on side, and pale on belly; dark blotch usually below eye; swollen belly of pregnant females with dark lateral blotch. **SIZE:** Maximum adult size 6.3 cm (2.5 in) TL, females larger than males. **RANGE:** The native range of eastern mosquitofish appears to be Atlantic coast drainages from Delaware Bay to Florida, extending into the Gulf of Mexico as far west as central Alabama. Eastern mosquitofish and the closely related western mosquitofish (*G. affinis*) have been introduced widely for mosquito control and have interbred, thus further confusing historic distribution records.

HABITAT AND HABITS: Eastern mosquitofish are found in shallow coastal streams, ponds, and ditches in fresh or brackish water. They typically dwell near the surface. In winter, eastern mosquitofish burrow into silty backwaters in streams or shallow brackish ponds. **OCCURRENCE IN THE CHESAPEAKE BAY:** Eastern mosquitofish are abundant in shallow sluggish coastal streams, ditches, and ponds throughout the Chesapeake Bay system.

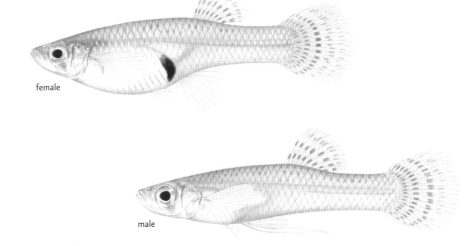

female

male

Eastern mosquitofish *Gambusia holbrooki*

REPRODUCTION: Eastern mosquitofish are livebearing and produce well-developed larvae 8–10 mm (0.3–0.4 in) in length that become mature at 4–6 weeks of age. Young are born from mid-April to September, with females capable of producing more than one brood per season. Probably few eastern mosquitofish live more than one year.

FOOD HABITS: Eastern mosquitofish eat insects, insect larvae, other small invertebrates, and plant material.

IMPORTANCE: Eastern mosquitofish have been cultured and introduced widely in the United States and globally for mosquito control. Unfortunately, many such exotic introductions have turned out badly, because mosquitofish outcompete and cause the decline of native fishes. In their native ecosystems, mosquitofish are important small forage species.

Fourspine stickleback - *Apeltes quadracus*
(Mitchill, 1815)

KEY FEATURES: No lateral bony scutes; typically three (sometimes four) large free dorsal spines, with another spine connected to soft dorsal fin. COLOR: Brownish green or black dorsally, silver ventrally, with dusky mottles scattered on body. When breeding, pelvic membranes reddish in males, orangish in females. SIZE: Maximum adult size 6.4 cm (2.5 in) TL. RANGE: Along the western Atlantic coast from Newfoundland to the Trent River system, North Carolina.

HABITAT AND HABITS: Primarily a nearshore species; however, as with threespine sticklebacks, these fish are often found in freshwaters. Fourspine sticklebacks have

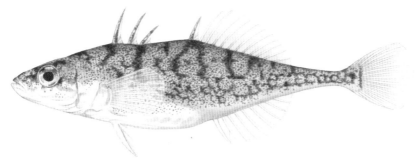

Fourspine stickleback *Apeltes quadracus*

the widest range of salinity tolerance of any North American species of stickleback. These fish are commonly associated with well-vegetated tidal creeks and brackish estuaries. Fourspine sticklebacks are not strong swimmers and are often seen hovering above aquatic vegetation. OCCURRENCE IN THE CHESAPEAKE BAY: Fourspine sticklebacks, year-round residents, are abundant throughout the entire Chesapeake Bay. During the summer, fourspine sticklebacks inhabit estuarine grass flats, and they retreat to deeper water in channels during winter.

REPRODUCTION: Fourspine sticklebacks differ from other sticklebacks in building a cup-shaped nest rather than a barrel-shaped one. The nest is ventilated by the fish pumping water through its gill covers. A single male may build and maintain several nests simultaneously. Spawning occurs in the Chesapeake Bay from late April to early May. The males usually have a one-year life span; some females may live to spawn again at age two.

FOOD HABITS: Fourspine sticklebacks feed along the bottom, primarily on diatoms, worms, and crustaceans, by sucking in the prey with a pipetting action.

IMPORTANCE: Of no recreational or commercial importance. Collected by seines along vegetated shorelines. Frequently collected with pipefishes and seahorses.

Threespine stickleback - *Gasterosteus aculeatus*
Linnaeus, 1758

KEY FEATURES: Few to many lateral bony plates; typically two large free dorsal spines with a third, smaller spine; free dorsal spines strong, the first inserting dorsal to pectoral base; second spine longest; third spine very short and connected to soft dorsal fin. COLOR: Dusky green dorsally, silver ventrally, with numerous black speckles on body. Males with reddish chins and bellies when trying to attract females. SIZE: Maximum adult size 10 cm (4 in) TL. RANGE: Known from both the eastern and western Atlantic coasts as well as the Pacific. In the western Atlantic, its range is from Labrador to the Chesapeake Bay.

HABITAT AND HABITS: Occur in freshwater streams and lakes, estuaries, and coastal seas. In the sea, confined to coastal waters. Inhabit shallow vegetated areas, usually over mud or sand. Threespine sticklebacks are often found in schools. The young are associated with drifting seaweed. OCCURRENCE IN THE CHESAPEAKE BAY: Threespine sticklebacks are visitors to the Chesapeake Bay during the winter and spring

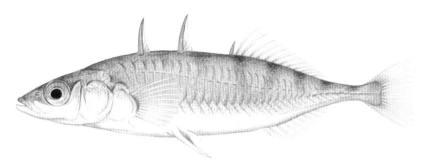

Threespine stickleback *Gasterosteus aculeatus*

(extending as far north as the Chester River) and are absent from the bay the remainder of the year.

REPRODUCTION: In our area, threespine sticklebacks are anadromous, ascending small western shore tributaries, such as along the James and York Rivers, to spawn. However, in the colder waters of the northern part of their range, they are strictly marine. Spawning occurs from late February to September. The spawning habits of sticklebacks are among the most extensively studied among fishes. Typically, the male constructs a nest from grasses and fibers bound together by a thread-like kidney secretion. The nest is tubular and contains a tunnel-like passageway. One or more females are then enticed into the nest to lay their eggs. The male guards and protects the eggs and young for a period of several weeks. The adult males are highly territorial and pugnacious, attacking other males or even other fishes who intrude near their nests. Egg laying occasionally occurs without nest building.

FOOD HABITS: Threespine sticklebacks are predators on small invertebrates such as worms, crustaceans, and larval and adult aquatic insects. These fishes have also been reported to feed on their own fry and eggs.

IMPORTANCE: Of no commercial or recreational importance.

Lined seahorse - *Hippocampus erectus* Perry, 1810

KEY FEATURES: Head shaped like that of a horse, placed nearly at a right angle to axis of body; tube-like snout; tail prehensile, long, and thin, pointed at tip, with no tail fin. **COLOR:** Variable, from yellow and brown to nearly black; dusky lines and spots on sides, head, and dorsal fin. **SIZE:** Maximum adult size 17 cm (6.7 in) TL. **RANGE:** Nova Scotia to Uruguay, including the Gulf of Mexico and the Caribbean Sea.

Lined seahorse *Hippocampus erectus*

HABITAT AND HABITS: At temperate latitudes, during the summer months, lined seahorses inhabit channels and flats; in winter they retreat to deeper waters. Lined seahorses are usually associated with seagrass beds or other structured habitats, such as sponges, pilings, and old crab pots, to which they cling with the prehensile tail. Both males and females have restricted home ranges, limited to only a few feet. **OCCURRENCE IN THE CHESAPEAKE BAY:** Lined seahorses are uncommon residents of the middle and lower Chesapeake Bay, extending as far north as the Patuxent River. They can be found at salinities as low as 9‰. In winter, they move into deeper channels or inshore ocean waters.

REPRODUCTION: Males and females often form pair bonds and, after an elaborate courtship, the female deposits eggs in the male's abdominal pouch, where they are fertilized. Males protect, aerate, and nourish the developing embryos for at least two weeks before hatching occurs inside the pouch at a size of about 4.0 mm (0.18 in) TL. The embryos are retained for a while longer before they are born at about 6.0 mm (0.24 in) TL. Males may carry about 100–1500 eggs, depending on body mass. Reproduction lasts from May to October.

FOOD HABITS: Lined seahorses are masters of camouflage and voracious ambush predators. They can change color in seconds to match their background and await potential prey items by clinging to vegetation or other holdfasts such as sponges, pilings, or ropes. Prey such as small crustaceans are ingested by sucking them through the long tubular snout.

IMPORTANCE: Lined seahorses have no commercial or recreational value, although they are sometimes collected by aquarium enthusiasts. Lined seahorses can be kept in an aquarium if live food (brine shrimp or zooplankton) is provided.

Dusky pipefish - *Syngnathus floridae*
(Jordan and Gilbert, 1882)

KEY FEATURES: Tube-like snout; 16–19 (typically 17) bony rings around trunk; dorsal fin short, having 26–34 rays and extending over 2 trunk rings and 4 or 5 tail rings. COLOR: Variable, near white to brownish green, with markings mostly tan to near black; snout typically with dusky lateral stripe; side and dorsum of posterior tail rings usually with characteristic pattern of irregular narrow brownish stripes; caudal fin brownish, often with pale margin. Dusky pipefish have been reported to change color to match their background. SIZE: Maximum adult size 26 cm (10.2 in) TL. RANGE: The Chesapeake Bay to the Gulf of Mexico and Caribbean as far south as Panama.

Dusky pipefish *Syngnathus floridae*

HABITAT AND HABITS: At temperate latitudes, dusky pipefish occur in estuaries from spring to autumn in vegetated or other structured habitats. In winter the species moves into deeper estuarine waters or out onto the inner continental shelf. OCCUR-

RENCE IN THE CHESAPEAKE BAY: Dusky pipefish are year-round residents of the middle to lower Chesapeake Bay, extending as far north as the Patuxent River. They are abundant in shallow water (1–3 m, or 3–10 ft) over grass flats during the summer, and they occupy deep channels in the winter.

REPRODUCTION: Dusky pipefish mature at age one and probably do not survive to age two. Dusky pipefish exhibit sex-role reversal and are polygamous, with an elaborate courtship behavior. Males typically are sexually mature at 14–15 cm (5.5–5.9 in) TL. The male has a specialized pouch on its belly into which the female deposits her eggs and where the eggs are fertilized and the embryos carried to term. Males may carry about 15–800 eggs, depending on body mass. Reproduction lasts from May to October, with a peak in July and August.

FOOD HABITS: Dusky pipefish feed mostly on small grass shrimp and other small crustaceans.

IMPORTANCE: Dusky pipefish have no commercial or recreational value.

Northern pipefish - *Syngnathus fuscus* Storer, 1839

KEY FEATURES: Tube-like snout, its depth about 20% of its length; bony rings around trunk numbering 18–21, typically 19; dorsal fin long, having 33–49 rays and extending over 4–5 trunk rings and 4–5 tail rings. COLOR: Variable, tan to dark brown and olive with darker markings; diagonal bar from eye to gill cover; 12–13 dark bands along side; tail fin with narrow light margin. SIZE: Maximum adult size 28 cm (11 in) TL; however, most individuals do not exceed 20 cm (8 in) TL. RANGE: Gulf of St. Lawrence to Florida.

Northern pipefish *Syngnathus fuscus*

HABITAT AND HABITS: At temperate latitudes, northern pipefish occur in estuaries from spring to autumn in vegetated or other structured habitats. In winter, this species moves into deeper estuarine waters or out onto the inner continental shelf. OCCURRENCE IN THE CHESAPEAKE BAY: Northern pipefish are common residents of seagrass beds in the middle and lower Chesapeake Bay during the warmer months.

They move into deeper channels in winter. They are the most common pipefish in the Chesapeake Bay.

REPRODUCTION: Northern pipefish mature at age one, and some individuals may survive to age two. Northern pipefish exhibit sex-role reversal and are polygamous, with an elaborate courtship behavior. The male has a specialized pouch on its belly into which the female deposits her eggs and where the eggs are fertilized and the embryos carried to term. Males may carry about 20–1,380 eggs, depending on body mass. Reproduction lasts from April to September, with a peak in June and July.

FOOD HABITS: Northern pipefish feed mostly on amphipods and other small crustaceans.

IMPORTANCE: Northern pipefish have no commercial or recreational value.

Chain pipefish - *Syngnathus louisianae* Günther, 1870

KEY FEATURES: Tube-like snout, its depth about 10% of its length; bony rings around trunk numbering 19–21, typically 20; dorsal fin having 33–42 rays and extending over 2–3 trunk rings and 5–6 tail rings. COLOR: Variable, tan to brown and olive, with a long series of chain-like diamond-shaped marks along lower side; brownish stripe on snout laterally; caudal fin brownish. SIZE: Maximum adult size 38 cm (15 in) TL. RANGE: Chain pipefish occur from New Jersey to Florida, the Gulf of Mexico, the Caribbean coast of Mexico, Jamaica, and Bermuda.

Chain pipefish *Syngnathus louisianae*

HABITAT AND HABITS: Chain pipefish are a warm-temperate to subtropical species that occupies seagrass beds and marshes and also floating mats of sargassum weed. They are most commonly found in depths of 10 m (33 ft) or less. OCCURRENCE IN THE CHESAPEAKE BAY: Chain pipefish have been infrequently reported from the Chesapeake Bay and are apparently rare summer visitors to the lower bay, extending as far north as the Potomac River.

REPRODUCTION: The age at maturity and maximum age of chain pipefish are unknown. Chain pipefish exhibit sex-role reversal. The male has a specialized pouch on its belly into which the female deposits her eggs and where the eggs are fertilized and the embryos carried to term. Males may carry about 900 eggs, depending on body mass.

FOOD HABITS: Chain pipefish feed mostly on amphipods and other small crustaceans.

IMPORTANCE: Chain pipefish have no commercial or recreational value.

Longhorn sculpin - *Myoxocephalus octodecemspinosus* (Mitchill, 1814)

Lumpfish - *Cyclopterus lumpus* Linnaeus, 1758

KEY FEATURES: The head of longhorn sculpin is bony and armed with numerous spines, some quite long. The body of lumpfish is globe-like, with thickened skin in which are embedded large horny tubercles; pelvic fins form a sucking disk; first dorsal fin with six to eight spines covered by thick skin in adults. COLOR: The head and back of longhorn sculpin are mottled brown, with several very dusky brown blotches. The pectoral, caudal, and soft dorsal fins have three brown streaks. Color in lumpfish is variable, with young being yellowish to greenish and adults bluish to brownish. Spawning males are reddish. SIZE: Maximum adult size of longhorn sculpin is about 45 cm (18 in) TL, while with lumpfish, it is about 60 cm (2 ft) TL. RANGE: Longhorn sculpin are found only in northwest Atlantic waters, and lumpfish are distributed along both coasts of the north Atlantic. Both species range southward along the North American coast as far as the Chesapeake Bay.

HABITAT AND HABITS: Longhorn sculpin are slow-moving bottom dwellers reported from depths as great as 190 m (630 ft). Lumpfish are typically associated with rocky bottoms in waters from a few meters to more than 300 m (1000 ft) deep. Adult lumpfish are quite sedentary, using the sucking disk to attach to rocks or other hard surfaces. OCCURRENCE IN THE CHESAPEAKE BAY: Both species are occasional visitors to the lower Chesapeake Bay in winter to early spring. Longhorn sculpin will enter river mouths and creeks but not freshwater.

REPRODUCTION: Longhorn sculpins are bottom spawners. Egg masses are deposited on the bottom and are often guarded by the male. Spawning lumpfish are often found inshore. Their demersal adhesive eggs are deposited in large clumps in compara-

Longhorn sculpin *Myoxocephalus octodecemspinosus*

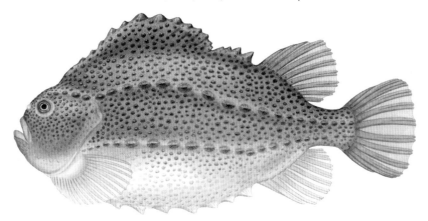

Lumpfish *Cyclopterus lumpus*

tively shallow water. Egg masses are guarded by the male and may include as many as 200,000 eggs.

FOOD HABITS: Longhorn sculpins are benthic feeders on worms, hydroids, shrimps, crabs, mussels, and small fishes. Lumpfish feed on ctenophores, jellyfish, crustaceans, polychaetes, and small fishes.

IMPORTANCE: Neither species has any commercial or recreational importance in the Chesapeake Bay region, but lumpfish eggs are marketed as caviar in more northern areas.

Northern searobin - *Prionotus carolinus* (Linnaeus, 1771)

KEY FEATURES: Dorsal fins separate, with 9 or 10 spinous rays in first dorsal fin and 13 or 14 soft rays in second dorsal fin; margin of caudal fin concave. **COLOR:** Body grayish or reddish brown above and pale below; gill membranes blackish, with five dusky saddle-like blotches along upper surface of body; first dorsal fin with dusky spot near outer edge of fin, typically between fourth and fifth spines, with white horizontal band under dusky spot; pectoral fin with spots on upper rays, with finger-like lower rays orange. **SIZE:** Maximum adult size about 45 cm (18 in) TL, common to about 20 cm (8 in) TL. **RANGE:** Along the western Atlantic coast from Nova Scotia to the east coast of Florida.

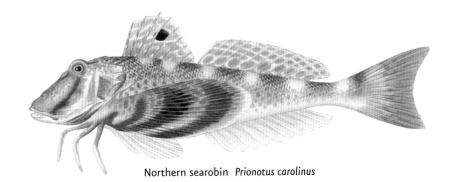

Northern searobin *Prionotus carolinus*

HABITAT AND HABITS: Northern searobins inhabit sandy bottoms, from shallow estuaries to the deeper waters at the edge of the continental shelf, but are found most commonly between 20 and 60 m (65–200 ft). Northern searobins, like striped searobins, use their modified pectoral rays to feel for food and also to stir up sand, weeds, and debris, thus dislodging prey items such as shrimps, crabs, squids, and small fishes. **OCCURRENCE IN THE CHESAPEAKE BAY:** Northern searobins are the most common searobin throughout the Chesapeake Bay and are abundant in the lower reaches, with records as far north as the Nanticoke River on the Eastern Shore and the Potomac River on the Western Shore. As with striped searobins, northern searobins are present in the deep flats and channels of the Chesapeake Bay from spring through early winter, retreating offshore or southward during the winter.

REPRODUCTION: Searobins are known to emit sounds, especially during the reproductive season. Some spawning by northern searobins may occur within the Chesapeake Bay during the summer; however, spawning is known to occur offshore of the bay during August and September. Searobin larvae commonly bury themselves in loose substrate.

FOOD HABITS: Crabs, worms, and fishes comprise much of the diet of northern searobins.

IMPORTANCE: Northern searobins are considered a nuisance by sport fishers because they steal bait and are unpleasant to handle. Commercial fishers view searobins as trash fish when caught incidentally in trawls and pound nets.

Striped searobin - *Prionotus evolans* (Linnaeus, 1766)

KEY FEATURES: Dorsal fins separate, with 9 or 10 spinous rays in first dorsal fin and 11–13 soft rays in second dorsal fin; margin of caudal fin truncate. **COLOR:** Body reddish to olive brown above and pale below; gill membranes pale or whitish; narrow black stripe extending along lateral line from head to caudal fin with second, incomplete stripe below it; first dorsal fin with dusky spot between fourth and sixth spines; pectoral fin orange brown, with numerous narrowly separated dusky streaks; leg-like pectoral filaments with brownish bars. **SIZE:** Maximum adult size approaches 50 cm (20 in) TL. **RANGE:** Along the western Atlantic coast from Nova Scotia to the east coast of Florida. Apparently rare north of Cape Cod.

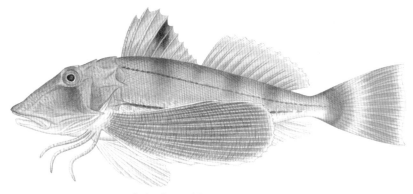

Striped searobin *Prionotus evolans*

HABITAT AND HABITS: Striped searobins are a temperate-subtropical species inhabiting sandy bottoms from inshore estuaries to about 180 m (595 ft), most commonly between 20 and 60 m (65–200 ft). **OCCURRENCE IN THE CHESAPEAKE BAY:** Striped searobins are regular visitors to the Chesapeake Bay from spring to early winter (more common in the lower bay than in the upper) in deep flats and along channel margins. During the winter they migrate offshore and to the south.

REPRODUCTION: Spawning occurs at night or in the evening within the Chesapeake Bay, from May through early July. Spawning also occurs offshore from the mouth of the Chesapeake Bay in August and September. As with northern searobins, striped searobins produce "grunting" or "barking" sounds during the spawning season, which are likely related to mate attraction and selection.

FOOD HABITS: Striped searobins feed primarily on shrimps, crabs, and fishes. As their mouths are larger than northern searobins', prey items are typically larger as well. Fishes comprise a larger percentage of the diet of striped searobins than of northern searobins.

IMPORTANCE: Larger than northern searobins, striped searobins are considered recreational game fish. They are often caught as bycatch in nets, but not in sufficient numbers to be commercially important.

Harvestfish - *Peprilus paru* (Linnaeus, 1758)

KEY FEATURES: Dorsal and anal fins greatly elongated anteriorly, with falcate margins, the longest rays much longer than head length; no pores near base of dorsal fin; pelvic fins lacking. COLOR: Pale blue to green dorsally, silvery with yellowish sheen ventrally. SIZE: Maximum adult size about 30 cm (1 ft) TL. RANGE: Along the western Atlantic coast from Maine to Uruguay, including the Gulf of Mexico and the eastern Caribbean Sea. Infrequently encountered north of the Chesapeake Bay.

HABITAT AND HABITS: Harvestfish are found in large schools in inshore and offshore waters, sometimes occurring in mixed schools with butterfish. Harvestfish are a pelagic species that displays a wide salinity tolerance (4‰ to full seawater) and is found over sand or mud bottoms. OCCURRENCE IN THE CHESAPEAKE BAY: Harvestfish, visitors to the Chesapeake Bay from April through October, are common in the lower and occasional in the middle Chesapeake Bay, extending as far north as the Chester River.

REPRODUCTION: Spawning occurs offshore in spring and early summer in the Chesapeake Bay region. Young-of-the-year harvestfish of about 25–30 mm (1.0–1.2 in) show up in the Chesapeake Bay in July and August. Juveniles are often found in shallow coastal waters in and around floating vegetation and jellyfishes.

FOOD HABITS: Juveniles feed primarily on plankton, whereas adults feed mainly on jellyfishes, small fishes, crustaceans, and worms.

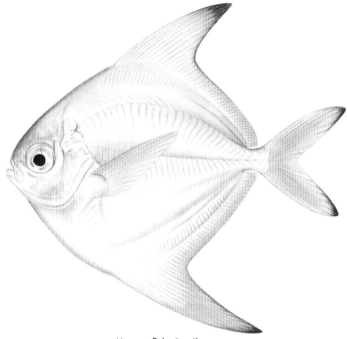

Harvestfish *Peprilus paru*

IMPORTANCE: Harvestfish are currently of minor commercial importance in the Chesapeake Bay; however, in former years catches were much larger. Between 1950 and 1965, the average annual catch was nearly 134 mt (295,375 lb), almost all from pound-net catches in Virginia waters. Since 1990, landings for the Virginia waters of the Chesapeake Bay are averaging less than 25 mt (55,125 lb) per year. Catches exhibit a single peak that usually occurs in May or June. Harvestfish are of limited recreational value, as they rarely take a baited hook.

Butterfish - *Peprilus triacanthus* (Peck, 1804)

KEY FEATURES: Dorsal and anal fins slightly elongate anteriorly, longest rays somewhat shorter than head length; series of well-developed conspicuous pores present near base of dorsal fin; pelvic fins lacking. **COLOR:** Pale blue to green along back, silvery belly with numerous irregular dusky spots on sides. **SIZE:** Maximum adult size 30 cm (1 ft) TL. **RANGE:** Along the western Atlantic coast from Nova Scotia to Florida and in the Gulf of Mexico.

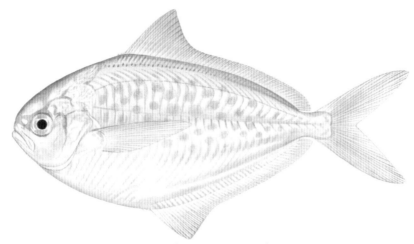

Butterfish *Peprilus triacanthus*

HABITAT AND HABITS: Butterfish form large schools in inshore and offshore waters, but typically at depths shallower than harvestfish do. Butterfish are pelagic and typically found over sand bottoms. **OCCURRENCE IN THE CHESAPEAKE BAY:** Butterfish occur in the Chesapeake Bay from March through November and are common to abundant in the lower bay and occasional in the upper bay, extending as far north as the Patapsco River near Baltimore. Within the bay, butterfish move northward in the spring, first appearing in Virginia waters in March, but they are not found above the Rappahannock River before May. All leave the Chesapeake Bay by December and overwinter offshore in deeper water (180–210 m, or 590–690 ft).

REPRODUCTION: Butterfish spawn offshore from May to July in the Chesapeake Bay region. After hatching, juveniles move from offshore surface waters to near-coastal waters, sometimes including bays and estuaries. They often hide from predators in masses of floating seaweed or among the tentacles of jellyfish.

FOOD HABITS: Juveniles feed primarily on plankton, whereas adults feed mainly on jellyfishes, small fishes, crustaceans, and worms.

IMPORTANCE: The flesh of butterfish is highly esteemed—however, as with harvest-fish, the recent landings of butterfish are significantly less than historical levels. Chesapeake Bay landings peaked at over 1,300 mt (2,904,900 lb) in 1965 and have generally declined since. Landings from Maryland and Virginia waters have not exceeded 100 mt (220,500 lb) since 1995 and, in this century, the annual commercial catch has only twice exceeded 50 mt (110,225 lb). Catches typically exhibit two peaks of abundance, the first usually occurring in April to May and the second in September to October. As they only rarely take a baited hook, butterfish do not constitute a recreational fishery.

Atlantic cutlassfish - *Trichiurus lepturus* Linnaeus, 1758

KEY FEATURES: Body ribbon-like, extremely elongate, and strongly compressed, tapering to a point; head long and compressed; mouth is large and contains strong fang-like teeth of unequal size, the largest ones with distinct barbs on posterior edges; lower jaw projecting, a short dermal process at its tip; pelvic fins absent. **COLOR:** Fresh specimens are metallic blue, whereas dead specimens are silvery gray. **SIZE:** Maximum adult size about 1.2 m (3.9 ft) TL, common to 50–100 cm (1.6–3.3 ft). **RANGE:** Worldwide in tropical and temperate waters. In the western Atlantic, the Atlantic cutlassfish ranges from Cape Cod to northern Argentina, including the Caribbean Sea and Gulf of Mexico.

Atlantic cutlassfish *Trichiurus lepturus*

HABITAT AND HABITS: These voracious predators are found throughout the water column and may occur to depths of 100 m (328 ft) or more but are often found in shallow coastal waters over muddy bottoms. **OCCURRENCE IN THE CHESAPEAKE BAY:** Atlantic cutlassfish are common members of the Chesapeake Bay fish fauna from spring to autumn, throughout mesohaline and polyhaline waters. They migrate offshore in autumn.

REPRODUCTION: In the western Atlantic, spawning occurs in late spring and summer over the continental shelf. The pelagic larvae and early juveniles are carried by currents into shallow waters, sometimes into estuaries. About half of Atlantic cutlassfish mature at the end of their first year of life, and all are mature at about two years.

FOOD HABITS: Atlantic cutlassfish feed on a wide variety of fishes, especially anchovies and sardines, as well as squids and shrimps. Juveniles feed on shrimps and small fishes, while adults are primarily piscivorous but will eat crustaceans and squids. Large adults typically feed near the surface during the day and spend the night near the bottom; juveniles and small adults feed in an opposite pattern to large adults.

IMPORTANCE: Although they readily take a bait, cutlassfish do not constitute a significant recreational resource in the Chesapeake Bay region, nor are they fished commercially. In the Chesapeake Bay, Atlantic cutlassfish are landed as bycatch in pound nets, bottom trawls, and beach seines.

Northern sand lance - *Ammodytes dubius*
Reinhardt, 1837

KEY FEATURES: Body very slender, slightly compressed; head long; snout sharply pointed; jaws lack teeth; caudal fin deeply forked; pelvic fins absent. COLOR: Olive, brown, or bluish green dorsally, with silvery sides and a white venter. SIZE: Maximum adult size to about 23 cm (9 in) TL, common to 10 cm (4 in) TL. RANGE: Labrador to Cape Hatteras.

Northern sand lance *Ammodytes dubius*

HABITAT AND HABITS: Northern sand lances are marine schooling fishes that will bury themselves in the bottom when threatened. They are found from the water's edge to depths of 70 m (about 225 ft). Northern sand lances inhabit substrates conducive to burrowing, such as sand bottoms with crushed shells or fine-gravelled bottoms. They will burrow into the substrate and settle into a resting position with their heads protruding at an angle. OCCURRENCE IN THE CHESAPEAKE BAY: Northern sand lances are rare visitors to the lower Chesapeake Bay from late summer to winter. A single specimen (12.5 cm, or 5 in, TL) collected off Back River near the mouth of the York River (VIMS 11605) is the only record we could locate from within the Chesapeake Bay.

REPRODUCTION: Spawning occurs from November to May at depths from 9 to 21 m (30–70 ft). Eggs are laid in or on the bottom substrate.

FOOD HABITS: Northern sand lances feed on planktonic organisms such as copepods and are preyed on by many commercially important species.

IMPORTANCE: Of no commercial or recreational importance in the Chesapeake Bay region, but they are an important forage species farther north.

African pompano - *Alectis ciliaris* (Bloch, 1787)

KEY FEATURES: Body compressed, with bluntly pointed snout; mouth large; first dorsal fin with 7 very short spines that become reabsorbed and embedded with age; spinous dorsal fin absent in fish longer than 15 cm (6 in) TL; first 5–7 rays of second dorsal and anal fins very long and filamentous in fish smaller than 40 cm (16 in) TL and sometimes in larger fish. **COLOR:** Most of body silvery blue to blue green dorsally, darkest on top of head and upper shoulder; remainder of body and head silvery. Juveniles have four or five dusky bars on body and black blotch near base of filamentous rays of second dorsal and anal fins. **SIZE:** Maximum adult size about 1.5 m (5 ft) in fork length (FL), common to about 1 m (3.3 ft). **RANGE:** Worldwide in tropical marine waters. In the western Atlantic, known from Massachusetts to Brazil, including throughout the Gulf of Mexico and Caribbean Sea.

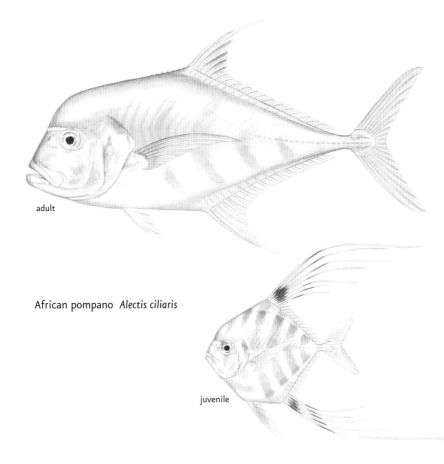

adult

African pompano *Alectis ciliaris*

juvenile

HABITAT AND HABITS: African pompano are a mostly solitary, strong-swimming species that is frequently found near the bottom, as deep as 60 m (197 ft). Juveniles and smaller adults of this species are among the most spectacular of jacks, possessing long delicate trailing filamentous dorsal- and anal-fin rays. Juveniles are usually pelagic and drifting; their streaming fin rays may help them resemble jellyfish and thus afford them a measure of protection from predation. OCCURRENCE IN THE CHESAPEAKE BAY: Juvenile African pompano are occasional visitors to the lower Chesapeake Bay during the summer and autumn but are rarely encountered in the middle bay, extending as far north as the Patuxent River (Western Shore) and the Choptank River (Eastern Shore). The largest specimen available to us from the Chesapeake Bay was collected from the mouth of the York River and measured 13 cm (5 in) FL.

REPRODUCTION: Little is known about the breeding habits of this species.

FOOD HABITS: The diet of African pompano consists of slow-swimming or sedentary crustaceans. Squids and fishes are also occasionally eaten.

IMPORTANCE: African pompano are of no local recreational or commercial importance in the Chesapeake Bay area but are esteemed as a food fish in regions where they are more abundant. The juveniles are sometimes taken in beach seines in the lower Chesapeake Bay.

Blue runner - *Caranx crysos* (Mitchill, 1815)

KEY FEATURES: Mouth large; eye moderate, with adipose eyelid; entire body covered with small scales; 46–56 prominent scutes on posterior part of lateral line. **COLOR:** Pale olive to bluish gray on back, silvery gray or golden on sides and belly. Juveniles possess 7 dusky bars on body. **SIZE:** Maximum adult size about 70 cm (2.3 ft) TL, common to about 35 cm (1.1 ft) TL. **RANGE:** Occurring throughout the Atlantic. In the western Atlantic, blue runners are known from Nova Scotia to southern Brazil, including the Gulf of Mexico and Caribbean Sea.

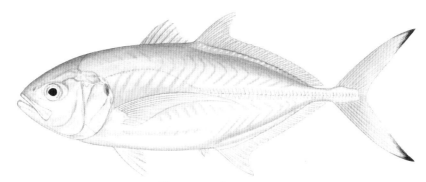

Blue runner *Caranx crysos*

HABITAT AND HABITS: Blue runners are one of the most common jacks in the western Atlantic and are found most often inshore, usually in large schools. **OCCURRENCE IN THE CHESAPEAKE BAY:** Blue runners are occasional to common visitors to the lower Chesapeake Bay during the summer and autumn and are sometimes encountered in the upper bay. The largest specimen recorded from the Chesapeake Bay was collected from Flag Ponds Nature Park, which is north of the mouth of the Patuxent River, and measured 24 cm (9.4 in) FL.

REPRODUCTION: Spawning occurs offshore and to the south of the Chesapeake Bay from January through August, with a peak in summer. The young have been collected offshore associated with jellyfish and other floating objects.

FOOD HABITS: Blue runners feed on fishes, crabs, and shrimps.

IMPORTANCE: Of no commercial or recreational importance in the Chesapeake Bay region.

Crevalle jack - *Caranx hippos* (Linnaeus, 1766)

KEY FEATURES: Mouth large; eye large, with adipose eyelid; chest lacks scales except for small patch of scales in front of pelvic fin. **COLOR:** Bluish green to dusky green along back, silvery white or golden on sides and belly; dusky oval blotch present on lower pectoral-fin rays; blackish spot on upper part of gill cover; caudal fin bright yellow while anal, pectoral, and second dorsal fins tinged with yellow; juveniles with five dusky bars on body. **SIZE:** Maximum adult size at least 100 cm (3.3 ft) FL, common to 60 cm (2.4 ft) FL. **RANGE:** Occurring throughout the Atlantic Ocean. In the western Atlantic, crevalle jacks are known from Nova Scotia to Uruguay, including the Gulf of Mexico and Caribbean Sea.

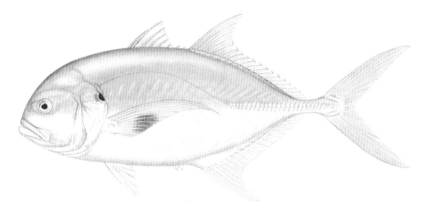

Crevalle jack *Caranx hippos*

HABITAT AND HABITS: Crevalle jacks typically travel in fast-moving, moderate- to large-size schools and often enter brackish water and ascend rivers. Larger fish are usually solitary and are most often found offshore. **OCCURRENCE IN THE CHESAPEAKE BAY:** Crevalle jacks occur occasionally to commonly in the lower Chesapeake Bay during the summer and autumn. Juvenile crevalle jacks have been collected between July and September near the head of the bay from the Elk and Sassafras Rivers.

REPRODUCTION: Spawning apparently occurs offshore from March through September. Little is known of the spawning and early life history of this species, and the majority of the spawning for crevalle jacks in the western Atlantic may take place south of U.S. waters. However, young-of-the-year of this species are fairly abundant in estuaries in the Mid-Atlantic Bight, indicating that these estuaries are used as nurseries.

FOOD HABITS: Like most large jacks, this species feeds primarily on fishes and a variety of invertebrates.

IMPORTANCE: Members of the genus *Caranx* are of considerable commercial importance worldwide but are generally held in low esteem as a food fish in the United States. Crevalle jacks are landed primarily with purse seines, gillnets, and handlines and in the Chesapeake Bay area by pound nets. However, the commercial catch in the Chesapeake Bay is insignificant. Anglers consider crevalle jacks good sport fish that will put up an energetic fight when taken with light tackle. Virginia waters of the Chesapeake Bay may yield a few thousand crevalle jacks to anglers each year.

Yellow jack - *Caranx bartholomaei* Cuvier, 1833
Horse-eye jack - *Caranx latus* Agassiz, 1831

KEY FEATURES: The body of yellow jacks is only moderately deep, whereas horse-eye jacks have a deep body. Yellow jacks have fewer lateral-line scutes than horse-eye jacks (22–28 vs. 32–39). The second dorsal fin of yellow jacks has 1 spine and 25–28 rays, while horse-eye jacks have 1 spine and 19–22 rays in the second dorsal fin. Yellow jacks also have more anal-fin rays than horse-eye jacks (21–24 vs. 16–18). COLOR: Both jacks are bluish gray dorsally and silvery white or golden on sides and belly, with yellow caudal fins. Some yellow jacks have all yellow fins with yellow blotches dorsally on the body. Horse-eye jacks sometimes have a small black spot on the upper part of the gill cover. SIZE: The maximum adult size of horse-eye jacks is about 80 cm (2.6 ft) TL, while yellow jacks approach 1 m (3.3 ft) TL. RANGE: Both species are found in the Gulf of Mexico and Caribbean Sea and extend as far south in the western Atlantic as Brazil. Yellow jacks occur as far north as Massachusetts, while horse-eye jacks only extend north as far as New Jersey. Horse-eye jacks are also found in the eastern Atlantic.

HABITAT AND HABITS: Both species are pelagic; however, horse-eye jacks most often occur near shore, whereas yellow jacks are more common offshore. Horse-eye jacks can be found in small schools around islands and offshore reefs and may enter brackish waters and ascend rivers. Juvenile horse-eye jacks are found along sandy beaches. OCCURRENCE IN THE CHESAPEAKE BAY: Both species are rare visitors to the lower Chesapeake Bay during summer and autumn. Records exist for both species from Northampton County on the lower Eastern Shore of Virginia.

REPRODUCTION: For both species, spawning apparently occurs offshore and south of U.S. waters from late winter / early spring to late summer / early autumn. Yellow jack young are found offshore in association with floating sargassum and jellyfish.

Yellow jack *Caranx bartholomaei*

Horse-eye jack *Caranx latus*

FOOD HABITS: Horse-eye jacks feed on small fishes and invertebrates, including shrimps, while yellow jacks feed primarily on fishes near the bottom.

IMPORTANCE: No commercial or recreational importance in the Chesapeake Bay region.

Atlantic bumper - *Chloroscombrus chrysurus*
(Linnaeus, 1766)

KEY FEATURES: Body ovate and extremely compressed; profile of belly much more convex than back; snout short; mouth upturned and small; caudal peduncle narrow, extended, and lacking keels. COLOR: Body and head dusky dorsally, sometimes with metallic blue sheen; sides and belly silvery; prominent black spot on upper part of caudal peduncle; fins, except for pelvic fins, yellowish. SIZE: Maximum adult size is about 26 cm (10 in) FL; common to 20 cm (8 in) FL. RANGE: Occur on both sides of the Atlantic. In the western Atlantic, known from Massachusetts to Uruguay, including coastal areas of the Gulf of Mexico and Caribbean Sea.

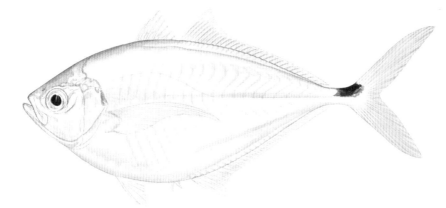

Atlantic bumper *Chloroscombrus chrysurus*

HABITAT AND HABITS: Atlantic bumper typically occur in schools in shallow coastal waters but are also found offshore to depths of 90 m (295 ft). They often emit a grunting sound when out of the water. Juveniles are often found offshore, occasionally in association with jellyfishes. OCCURRENCE IN THE CHESAPEAKE BAY: Atlantic bumper are uncommon visitors to the lower Chesapeake Bay during autumn.

REPRODUCTION: The probable spawning period for Atlantic bumper is during the spring and summer along the southeastern U.S. coast.

FOOD HABITS: Atlantic bumper feed on fishes, squids, and zooplankton.

IMPORTANCE: Atlantic bumper are of no recreational importance but are occasionally caught on hook and line.

Leatherjack - *Oligoplites saurus* (Bloch and Schneider, 1801)

KEY FEATURES: First dorsal fin with 4–6 spines, typically 5; second dorsal fin high anteriorly and long, with 1 spine and 19–21 soft rays; anal fin with 3 spines and 19–22 soft rays, first 2 anal-fin spines well in advance of third. **COLOR:** Metallic bluish to greenish along back, silvery white on sides and belly; sometimes with seven or eight irregular broken silvery bars and white interspaces along middle of sides; caudal fin amber or yellow. **SIZE:** Maximum adult size approaches 30 cm (1 ft) FL; common to about 27 cm (10.6 in) FL. **RANGE:** From Massachusetts to Brazil, including coastal areas of the entire Gulf of Mexico and Caribbean Sea.

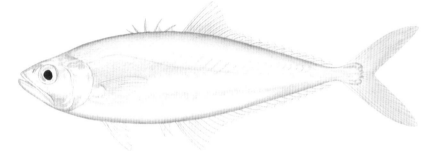

Leatherjack *Oligoplites saurus*

HABITAT AND HABITS: Leatherjacks are schooling fast-moving fish that frequent inshore habitats, usually along sandy beaches and in bays and inlets. They tolerate low salinities and are found more often in turbid rather than clear water. **OCCURRENCE IN THE CHESAPEAKE BAY:** Leatherjacks are rare visitors to the lower Chesapeake Bay during summer and autumn. The largest specimen available to us from the Chesapeake Bay was 23 cm (9 in) FL.

REPRODUCTION: Spawning occurs in shallow inshore waters from early spring to mid-summer. Juveniles have been observed floating at the surface with tail bent and head down, suggesting a floating leaf.

FOOD HABITS: Adult leatherjacks feed on fishes and crustaceans. Reports exist of juvenile leatherjacks feeding on fish scales and serving as "parasite pickers" for other fishes. Juveniles have incisor-like outer teeth that enable this type of feeding behavior. As juveniles mature, their teeth become conical, and their diet changes.

IMPORTANCE: Leatherjacks are considered excellent fighting fish on ultralight tackle. The two detached anal-fin spines of leatherjacks are purportedly connected to toxic glands, and in Florida this fish has been called the "stinging jack."

Bigeye scad - *Selar crumenophthalmus* (Bloch, 1793)
Rough scad - *Trachurus lathami* Nichols, 1920

KEY FEATURES: Rough scad differ from bigeye scad in being more elongate and less stout. Rough scad have large, hardened scales (scutes) covering the entire lateral line, whereas bigeye scad have large scutes only on the posterior part of the lateral line. Bigeye scad have fewer rays (24–27) in the second dorsal fin than rough scad (28–34). **COLOR:** Coloration is similar in both bigeye and rough scad: dorsal surface of the body and head metallic blue or bluish green, rest of the body and head silvery or whitish, with a small blackish spot near the upper margin of the gill cover. In bigeye scad, there is often a pale yellow stripe extending from the gill cover to the caudal fin. **SIZE:** The maximum adult size of bigeye scad (27 cm, or 11 in, SL) is slightly less than that of rough scad (33 cm, or 1.1 ft, SL). **RANGE:** Bigeye scad are found world-

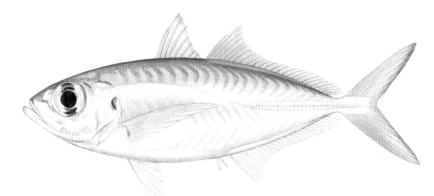

Bigeye scad *Selar crumenophthalmus*

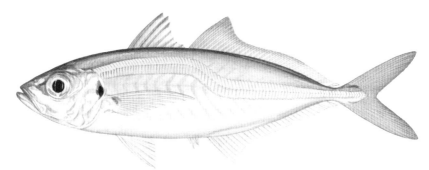

Rough scad *Trachurus lathami*

wide in tropical and subtropical marine waters. In the western Atlantic, bigeye scad are known from Nova Scotia to Brazil, including throughout the Gulf of Mexico and Caribbean Sea. Rough scad are known only from the Gulf of Maine to northern Argentina, including coastal areas of the Gulf of Mexico and Caribbean Sea.

HABITAT AND HABITS: Bigeye scad are found in small or large schools, mainly inshore, but they are known from all depths to about 170 m (560 ft). Rough scad are also a schooling species, and they occur from surface waters to depths of about 90 m (295 ft). OCCURRENCE IN THE CHESAPEAKE BAY: Bigeye scad are occasional to common visitors to the lower bay during the summer and autumn but are rarely encountered in the middle bay, extending as far north as the Patuxent River. Rough scad are occasional summertime visitors to the lower bay and are rare to occasional in the upper bay.

REPRODUCTION: Rough scad are believed to spawn offshore from April to June; juveniles have been collected offshore associated with jellyfishes. Little is known about the breeding biology of bigeye scad.

FOOD HABITS: Bigeye scad feed on small shrimp and benthic invertebrates when inshore and on zooplankton and fish larvae when offshore. Rough scad feed on small invertebrates.

IMPORTANCE: Neither bigeye scad nor rough scad is of any commercial or recreational importance. They are taken mainly as bycatch with trawls and seines and also occasionally by hook and line.

Atlantic moonfish - *Selene setapinnis* (Mitchill, 1815)

KEY FEATURES: Front of head slightly concave in profile, rising nearly vertically, then forming rounded edge over eyes; body short, deep, and extremely compressed; snout bluntly pointed and mouth large; scales very small and difficult to see, most of lower body with scales; first dorsal fin with 8 short spines; in juveniles less than 6 cm (2.5 in) FL, first 4 dorsal-fin spines filamentous. **COLOR:** Body and head silvery, sometimes with metallic blue-green sheen on upper body. Juveniles also silvery, with blackish, oval spot above lateral line under posterior part of second dorsal fin. **SIZE:** Maximum adult size is about 33 cm (1.1 ft) FL, common to about 24 cm (9.5 in) FL. **RANGE:** Known from Nova Scotia to Argentina, including coastal areas of the Gulf of Mexico and Caribbean Sea. However, not common north of the Chesapeake Bay.

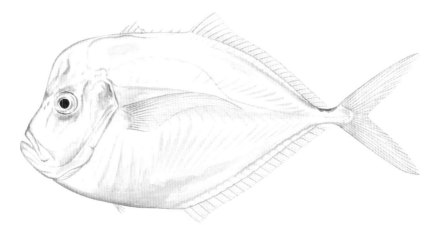

Atlantic moonfish *Selene setapinnis*

HABITAT AND HABITS: Atlantic moonfish are schooling fish, usually found near the bottom in inshore waters, but they may occur to depths of over 50 m (165 ft). Juveniles are frequently found in bays and river mouths. **OCCURRENCE IN THE CHESAPEAKE BAY:** Atlantic moonfish are regular visitors to the lower Chesapeake Bay during summer and autumn but are rarely encountered in the middle bay, extending as far north as the Patuxent River. The largest specimen available to us from the Chesapeake Bay was 7.8 cm (3 in) TL.

REPRODUCTION: Little is known about the breeding habits of Atlantic moonfish.

FOOD HABITS: Atlantic moonfish feed on small crustaceans and fishes and are reported to be nocturnal feeders.

IMPORTANCE: Said to be tasty, though providing little flesh. Where abundant, this species is fished commercially with trawls, seines, and pound nets. Atlantic moonfish are of little recreational importance.

Lookdown - *Selene vomer* (Linnaeus, 1758)

KEY FEATURES: Body short, deep, extremely compressed; head very deep, with front of head slightly concave in profile, rising nearly vertically, then forming sharp angle above eyes; snout bluntly pointed; first dorsal fin with 8 spines, first and second very long and filamentous in juveniles; second dorsal fin with 1 spine and 20–23 soft rays; spine of second dorsal and anal fins notably elongate. COLOR: Body and head silvery or golden, sometimes with metallic blue sheen on upper body. Juveniles silvery, with faint dusky bars on sides and grayish-black elongate dorsal and pelvic fins. SIZE: Maximum adult size is approximately 40 cm (1.3 ft) FL; common to 24 cm (9.5 in) FL. RANGE: Known from Maine to Uruguay, including coastal areas of the Gulf of Mexico and Caribbean Sea. Lookdowns are more common south of the Chesapeake Bay.

HABITAT AND HABITS: This species frequents hard or sandy bottoms around pilings and bridges in shallow coastal waters, usually in small schools near the bottom. OCCURRENCE IN THE CHESAPEAKE BAY: Juvenile lookdowns are common visitors to the lower Chesapeake Bay during summer and autumn and are occasionally encountered in the upper bay, extending as far north as the Bay Bridge near the mouth of the Chester River. The largest specimen available to us from the Chesapeake Bay was 14 cm (5.5 in) FL.

REPRODUCTION: Little is known about the breeding habits of this species.

FOOD HABITS: Lookdowns feed on small crustaceans, worms, and fishes.

IMPORTANCE: Where abundant, lookdowns are landed as commercial bycatch in trawls, seines, and pound nets. Lookdowns are of minimal recreational importance.

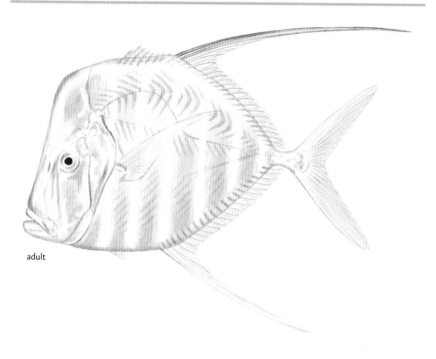

adult

Lookdown *Selene vomer*

juvenile

Greater amberjack - *Seriola dumerili* (Risso, 1810)

KEY FEATURES: First dorsal fin with 7 spines, last spine reduced and skin-covered in fish over about 55 cm (1.8 ft) FL; second dorsal fin with 1 spine and 29–34 soft rays; anal fin with 3 spines and 18–22 soft rays; first 2 anal-fin spines in advance of third, but may be skin-covered or recessed in large adults; groove present along top and bottom of caudal peduncle just before caudal fin. COLOR: Brownish or olivaceous along back, silvery white on sides and belly; head with dusky stripe originating at upper jaw and extending through eye to first dorsal fin; some individuals have a longitudinal amber stripe that extends from eye to tail; juveniles with six dusky bars on body. SIZE: This species is the largest of the jacks. Maximum adult size is about 1.9 m (6.2 ft) FL; common to 70 cm (2.3 ft) FL. RANGE: Worldwide in tropical and temperate seas, except the eastern Pacific. The western Atlantic distribution is from Nova Scotia to Brazil, including the entire Gulf of Mexico and Caribbean Sea.

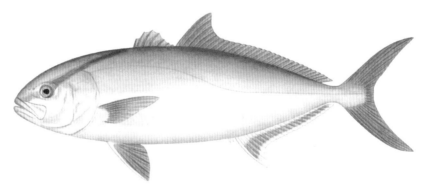

Greater amberjack *Seriola dumerili*

HABITAT AND HABITS: Greater amberjacks utilize the entire water column, with juvenile fish often found in waters less than 10 m (33 ft) deep. Adult fish frequent deeper waters, sometimes exceeding 100 m (328 ft). They are often found in small- to moderate-size schools but can be solitary. Juveniles are known to associate with floating mats of seaweed. OCCURRENCE IN THE CHESAPEAKE BAY: During summer, greater amberjacks are occasional visitors to the lower Chesapeake Bay, where they concentrate around reefs, rock outcrops, and wrecks. Common in most summers just offshore of the Chesapeake Light Tower, located in the Atlantic near the bay mouth.

REPRODUCTION: Greater amberjacks spawn offshore from March through July, with a peak in May or June.

FOOD HABITS: Fishes, crabs, and squids comprise the diet of greater amberjacks.

IMPORTANCE: The recreational catch in Virginia waters varies each year from a few hundred to a few thousand fish. Greater amberjacks are considered excellent fighting fish and will take live, dead, or artificial bait fished on the bottom or trolled. The flesh of greater amberjacks is slightly oily but has a mild and excellent flavor. In large fish the more posterior muscles, especially those of the caudal peduncle, are often infested with long thin parasitic worms. Although this parasite is not transmittable to humans, most people will want to discard the infested portion. An alternative is to remove the worms individually from the fish; they usually pull free from the raw flesh rather easily.

Almaco jack - *Seriola rivoliana* Valenciennes, 1833

KEY FEATURES: First dorsal fin with 7 spines, the last reduced and skin-covered in fish over about 55 cm (1.8 ft) FL; second dorsal fin with 27–33 soft rays, first few fin rays elevated; transverse groove dorsally and ventrally on caudal peduncle just before caudal fin; anal-fin base considerably shorter than second dorsal fin, anal fin with 3 spines and 18–22 soft rays, first 2 anal-fin spines in advance of third, but may be skin-covered or recessed in large adults, first few anal-fin rays elevated. **COLOR:** Head and body uniformly brown or bluish green; pale brass or lavender on sides and belly; frequently a faint dusky stripe extending from eye to dorsal-fin origin as well as a faint yellowish longitudinal stripe from eye to caudal fin; fins often blackish; juveniles with five irregular bars on body. **SIZE:** Maximum adult size about 97 cm (3.2 ft) FL, common to about 55 cm (1.8 ft) FL. **RANGE:** Circumtropical in marine waters,

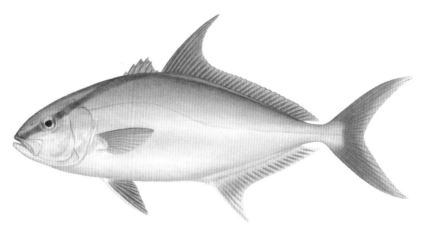

Almaco jack *Seriola rivoliana*

entering temperate waters in some areas. In the western Atlantic, almaco jacks are found from Cape Cod to Argentina, including the Gulf of Mexico and Caribbean Sea.

HABITAT AND HABITS: Almaco jacks typically frequent offshore waters and live in small groups, typically around wrecks, buoys, and reefs. Adult almaco jacks occur throughout the water column from 5–160 m (16–525 ft) in depth. OCCURRENCE IN THE CHESAPEAKE BAY: This species is a rare to occasional visitor during summer and autumn in the lower Chesapeake Bay, extending as far north as Mobjack Bay near the York River mouth.

REPRODUCTION: Little is known about this species, but spawning is believed to occur offshore in spring through fall. The pelagic juveniles are found offshore under floating sargassum and debris.

FOOD HABITS: Almaco jacks are fast-swimming predators that feed mainly on fishes but will also eat invertebrates such as squids.

IMPORTANCE: Almaco jacks are occasionally taken by commercial fisheries; however, there is no directed commercial fishery for this species. Most catches are by anglers, who consider almaco jacks to be excellent fighting fish. Almaco jacks will take live, dead, or artificial bait fished on the bottom or trolled.

Banded rudderfish - *Seriola zonata* (Mitchill, 1815)

KEY FEATURES: First dorsal fin with 8 spines, the last reduced and skin-covered in fish over about 55 cm (1.8 ft) FL; second dorsal fin with 1 spine and 33–40 rays; caudal peduncle narrow and extended, with transverse groove dorsally and ventrally; anal fin with 3 spines and 19–21 soft rays (about half as many as second dorsal fin); first 2 anal-fin spines in advance of third, but may be skin-covered or recessed in large adults. COLOR: Bluish gray dorsally, silvery white on sides and belly; tips of caudal-fin lobes are white; dusky stripe from eye to dorsal-fin origin present in juveniles and may be present in adults. Juveniles (to about 30 cm, or 1 ft, FL) with six blackish bars on body and fins as well as black stripe from eye to origin of dorsal fin. Juvenile banded rudderfish resemble pilotfish. SIZE: Maximum adult size is about 80 cm (2.6 ft) FL; common to 47 cm (1.5 ft) FL. RANGE: Known from Maine to Brazil, including coastal areas of the Gulf of Mexico and Caribbean Sea.

HABITAT AND HABITS: Banded rudderfish are most common in coastal waters over the continental shelf. Adults are free swimming, often near the bottom. Juveniles

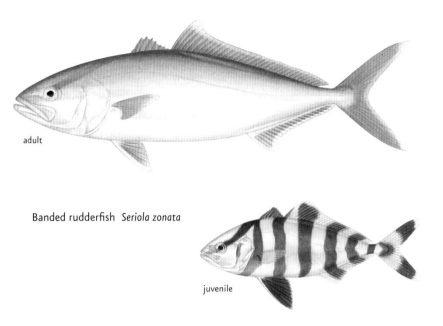

adult

Banded rudderfish *Seriola zonata*

juvenile

migrate northward along the U.S. Atlantic seaboard during the summer and are frequently found under or around jellyfishes, drifting seaweed, or larger fish. OCCURRENCE IN THE CHESAPEAKE BAY: This species is an occasional visitor to the lower Chesapeake Bay during the summer, extending as far north as the York River on the Western Shore and Cape Charles on the Eastern Shore. Banded rudderfish that enter the bay are typically no larger than 26 cm (10 in) FL.

REPRODUCTION: Spawning occurs year-round (with a possible summer hiatus) in offshore waters south of Cape Hatteras.

FOOD HABITS: Banded rudderfish feed on fishes and shrimps.

IMPORTANCE: Banded rudderfish are caught incidentally by commercial and recreational fisheries. There is no directed fishery anywhere in the range of this species. Banded rudderfish are taken in traps and pound nets, and with rod and reel, by trolling or bottom fishing.

Florida pompano - *Trachinotus carolinus* (Linnaeus, 1766)

KEY FEATURES: First dorsal fin with 6 spines; second dorsal fin with 1 spine and 22–27 (typically 23–25) soft rays; anal fin with 3 spines and 20–24 (typically 21 or 22) soft rays; first 2 anal-fin spines in advance of third. **COLOR:** Metallic green to bluish green on upper body and head, silvery on side, and whitish to yellowish on belly. **SIZE:** Maximum adult size to at least 60 cm (2 ft) FL, but rarely over 40 cm (1.3 ft) FL. **RANGE:** From Massachusetts to Brazil, including coastal areas of the entire Gulf of Mexico and Caribbean Sea.

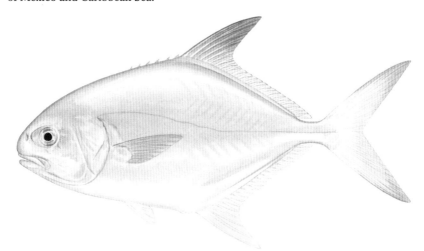

Florida pompano *Trachinotus carolinus*

HABITAT AND HABITS: Florida pompano are usually found in shallow waters along sandy beaches, particularly in the surf zone, around inlets, and in brackish-water bays, often moving with the tide. **OCCURRENCE IN THE CHESAPEAKE BAY:** Florida pompano are common visitors to the lower and middle Chesapeake Bay during summer and autumn, extending as far north as the Patuxent River. Small to large schools migrate north in summer along the U.S. Atlantic seaboard and are most abundant in the bay in July and August.

REPRODUCTION: Spawning apparently occurs offshore in late spring to summer. After moving inshore, juveniles are most abundant along low-energy beaches, where large schools have been reported to form.

FOOD HABITS: Florida pompano feed on clams, shrimps, crabs, and mussels.

IMPORTANCE: Florida pompano are renowned as a gourmet food item that commands a high price per pound but are of minor commercial importance in the Chesapeake Bay region. This species is also held in high regard as a game fish for its fast strikes and runs. Florida pompano are caught from the surf, off piers, and in boats on shallow flats by using jigs baited with crabs or mole crabs. In Virginia's bay waters, the number of Florida pompano caught annually by anglers can vary from a few hundred to more than 15,000.

Permit - *Trachinotus falcatus* (Linnaeus, 1758)
Palometa - *Trachinotus goodei* Jordan and Evermann, 1896

KEY FEATURES: In palometa, the longest dorsal- and anal-fin rays reach well beyond the caudal-fin base, whereas in permit, the longest dorsal- and anal-fin rays do not reach the caudal-fin base. In addition, the teeth in permit are resorbed by 20 cm (8 in) FL, while palometa have teeth at all sizes. COLOR: Palometa typically have four narrow vertical bars and two spots along the lateral line between the head and caudal peduncle. Permit have neither bars nor spots but may have a dusky to blackish blotch on the side. SIZE: The maximum adult size of permit is about 1.1 m (3.6 ft) FL, whereas in palometa, the maximum adult size is about 51 cm (1.7 ft) FL. RANGE: Both permit and palometa range from Massachusetts to Brazil, including coastal areas of the entire Gulf of Mexico and Caribbean Sea. Palometa extend southward to Argentina.

HABITAT AND HABITS: Permit occur from surface to bottom, in shallow water to depths greater than 35 m (115 ft). They frequent channels, holes, sandy flats, reefs, and at times mud bottoms. Adult permit tend to be solitary or travel in small schools, while juveniles can be found in large schools, especially during summer in the surf zone along sandy beaches. Juvenile permit can also tolerate brackish waters. Palometa are found around reefs and rocky areas and prefer high-salinity waters. On rare occasions palometa move into inlets and bays. OCCURRENCE IN THE CHESAPEAKE BAY: Permit are occasional visitors to the lower Chesapeake Bay during the summer and autumn. This species is rarely encountered in the middle portion of the bay, extending as far north as the Potomac River. Palometa are rare visitors to the lower bay.

REPRODUCTION: Little is known of the spawning habits and early life history of these species. Spawning probably occurs offshore in Gulf Stream waters.

FOOD HABITS: Permit feed primarily on mollusks, crustaceans, and small fishes, whereas palometa feed on fishes and small invertebrates. Permit will often follow in the wake of large stingrays, to forage on invertebrates stirred up by the stingrays while they feed.

IMPORTANCE: Due to their rarity, permit and palometa are of no commercial or recreational importance in the Chesapeake Bay region. In Florida and the tropics, permit are one of the most sought-after game fishes.

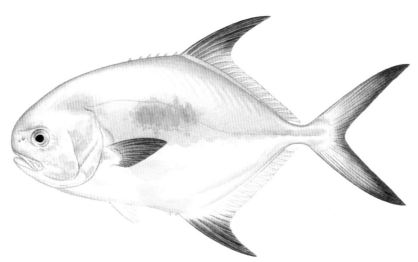

Permit *Trachinotus falcatus*

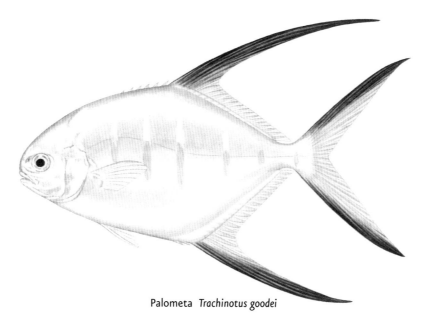

Palometa *Trachinotus goodei*

Cobia - *Rachycentron canadum* (Linnaeus, 1766)

KEY FEATURES: First dorsal fin with seven to nine (typically eight) short, stout isolated spines that fit in a groove when fin is depressed. **COLOR:** Blackish along back, brown on sides, and yellowish to dusky on belly; sides with two sharply defined white stripes that sandwich a broad dusky lateral stripe running from tip of snout to base of tail, uppermost white stripe less distinct in large fish. **SIZE:** Maximum adult size about 2 m (6.6 ft) TL, but more common at about half that size. **RANGE:** Cobia, the only member of the family Rachycentridae, are distributed worldwide in tropical to warm-temperate coastal waters. In the western Atlantic, they are known from Massachusetts to Argentina, including the Gulf of Mexico and Caribbean Sea.

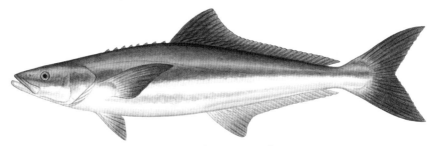

Cobia *Rachycentron canadum*

HABITAT AND HABITS: Cobia are most often found in open water around buoys, pilings, or floating objects. Cobia are most frequently solitary but occasionally found in small groups. They sometimes associate with sharks, rays, and pilotfish. **OCCURRENCE IN THE CHESAPEAKE BAY:** These summer visitors are common in the lower Chesapeake Bay and occasionally encountered in mid-bay waters. The northernmost record in the bay is from Breezy Point, which is located north of the Patuxent River mouth. Cobia enter the bay in late May or early June and migrate out of the bay and move south by mid-October.

REPRODUCTION: Spawning occurs from mid-June to mid-August near the Chesapeake Bay mouth or just offshore, where cobia form aggregations. Cobia are pelagic spawners, with both the eggs and larvae becoming part of the plankton.

FOOD HABITS: Cobia are opportunistic feeders. Although cobia eat fishes and squids, the bulk of their diet is crabs and shrimps; hence the colloquial name is "crabeater" in some areas.

IMPORTANCE: Caught incidentally throughout most of their range, because of their solitary habits. Commercial landings in the region amounted to about 2.9 mt (6,324 lb) in 2008, with almost all of the catch coming from Virginia waters. Cobia are highly prized by sport fishers. Cobia are considered excellent eating fish and rugged fighters that will make determined runs and leaps when hooked. Anglers in the lower Chesapeake Bay use either live or artificial bait to fish for cobia around buoys, towers, bridges, and other open-water structures. The Chesapeake Bay record for cobia was taken off York Spit Reef in 2006 and weighed 49.5 kg (109 lb). Due to its rapid growth and good flesh quality, aquaculture of cobia is being pursued by numerous countries, including the United States.

Sharksucker - *Echeneis naucrates* Linnaeus, 1758

KEY FEATURES: Body spindle-shaped, with broad, flattened head; on top of head, a large oval-shaped sucking disk that enables sharksuckers to attach themselves to other marine animals. COLOR: Back dusky brown with pale brown sides; dusky longitudinal stripe bounded by white stripes extends from mouth through eye to caudal fin; in juveniles, dorsal and anal fins are whitish at tips and caudal fin is whitish at top and bottom. SIZE: Maximum adult size is about 1.1 m (3.6 ft) TL. RANGE: Worldwide in tropical and temperate seas. Sharksuckers are the most common remora in the western Atlantic and are known from Massachusetts to Uruguay, including the Gulf of Mexico.

Sharksucker *Echeneis naucrates*

HABITAT AND HABITS: Sharksuckers are often present in shallow inshore brackish areas as well as around coral reefs. They are found at depths as great as 50 m (164 ft). Sharksuckers are often seen swimming free in small groups. Using their sucking disk, sharksuckers attach themselves onto either the body or the gill area of a variety of hosts, including sharks, rays, dolphins, sea turtles, and whales. While attached to hosts, the hosts are not harmed, but sharksuckers benefit immensely. Not only do sharksuckers obtain their food via the hosts, but they also save energy by "hitchhiking" a ride on the hosts. Sharksuckers are also known to attach to ships, buoys, floating objects, and even scuba divers. OCCURRENCE IN THE CHESAPEAKE BAY:

Sharksuckers are occasional summertime visitors to the Chesapeake Bay, reaching as far north as the mouth of the Chester River.

REPRODUCTION: Spawning occurs during June and July in the mid-Atlantic region. Sperm and eggs are released and fertilized externally. During development of the newly hatched fish, the sucking disk begins to form. After the fish have been living freely for about a year, the sucking disk is fully formed, and juvenile sharksuckers are able to attach to hosts.

FOOD HABITS: Sharksuckers feed on bits of food discarded or lost by their hosts as well as small parasites cleaned from the hosts' skin. This diet is augmented with free-living crabs, squids, and small fishes. Juvenile sharksuckers sometimes set up cleaning stations where they pick parasites off other fishes.

IMPORTANCE: Of no commercial or recreational value but occasionally caught on hook and line. There are reports of hooked sharksuckers quickly attaching themselves to large fish.

Little tunny - *Euthynnus alletteratus* (Rafinesque, 1810)

KEY FEATURES: Spinous dorsal fin high anteriorly, with concave margin and 15–16 spines; 8 dorsal and 7 anal finlets. COLOR: Greenish to dark blue dorsally, posterior upper body with wavy reticulated lines; chest below pectoral fin with four to five dusky spots; lower sides and belly silvery white. SIZE: Maximum adult size 1 m (3.3 ft) FL; common to 75 cm (2.5 ft) FL. RANGE: Found in tropical and subtropical wa-

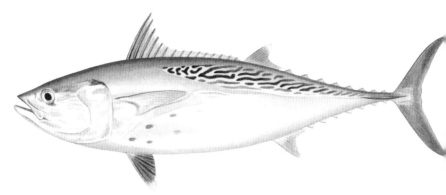

Little tunny *Euthynnus alletteratus*

ters of the Atlantic Ocean. In the western Atlantic, little tunny are known from New England to Brazil, including the Gulf of Mexico and Caribbean Sea.

HABITAT AND HABITS: Little tunny are an inshore schooling species that occurs in near-surface waters. During the seasonal north-south migrations the schools can be very large, consisting of thousands of fish, and often marked by the presence of diving birds that are feeding on the same food source. **OCCURRENCE IN THE CHESAPEAKE BAY:** Little tunny are occasional visitors to the Chesapeake Bay from late spring to fall.

REPRODUCTION: Spawning occurs along the edge of the continental shelf from spring through fall. Little tunny live about five years.

FOOD HABITS: Little tunny are opportunistic feeders on crustaceans, fishes, and squids.

IMPORTANCE: Anglers pursue little tunny by trolling or casting lures to schooling fish at the surface. The bloody flesh limits the popularity of little tunny as a food fish, and they are often used as bait for billfishes. Those who take the trouble to bleed and ice the fish immediately upon landing find little tunny highly acceptable, especially if the dark lateral muscle mass is discarded.

Atlantic bonito - *Sarda sarda* (Bloch, 1793)

KEY FEATURES: First dorsal fin low, with straight margin, 20–23 dorsal-fin spines, 7–9 dorsal finlets. **COLOR:** Steel blue above, with upper sides marked with 7–12 oblique stripes extending below lateral line; no spots on chest; lower sides and belly silvery. **SIZE:** Maximum adult size 90 cm (3 ft) FL; common to 50 cm (1.6 ft) FL.

Atlantic bonito *Sarda sarda*

RANGE: Found in tropical and temperate waters of the Atlantic Ocean. In the western Atlantic, Atlantic bonito are known from Massachusetts to Florida and the northern Gulf of Mexico but apparently are absent in most of the Caribbean. They are also known from Colombia and Venezuela as well as northern Argentina.

HABITAT AND HABITS: This pelagic migratory species often schools near the surface in nearshore waters. Atlantic bonito tolerate temperatures of 12–27°C (54–81°F) and salinities of 14–39‰ and occasionally enter estuaries. **OCCURRENCE IN THE CHESAPEAKE BAY:** Atlantic bonito are occasional visitors to the Chesapeake Bay from spring to fall.

REPRODUCTION: Spawning is in June and July in the northwestern Atlantic; however, spawning occurs during the winter months in waters south of Cape Hatteras.

FOOD HABITS: Atlantic bonito feed on small, schooling fishes (including other bonito), squids, and shrimps.

IMPORTANCE: Sought by commercial fisheries over their entire range, where abundant. In the western Atlantic, most commercial landings come from coastal waters from Mexico to Venezuela. A small fishery also exists in the Gulf of Maine from June through October. Sport fishers pursue this species by trolling or casting lures to schooling fish at the surface. The eating quality of Atlantic bonito is reported to be excellent.

Atlantic chub mackerel - *Scomber colias* Gmelin, 1789

KEY FEATURES: Front and back portions of eye covered by adipose eyelid; first dorsal fin with 8–10 slender spines; 5 finlets following second dorsal and anal fins. **COLOR:** Steel blue dorsally with faint dusky lines that zig-zag and undulate; silvery yellow ventrally, with numerous dusky rounded blotches or wavy broken lines. **SIZE:** Maximum adult size to about 50 cm (1.6 ft) FL; common to 30 cm (1 ft) FL. **RANGE:** Inhabiting warm and temperate waters of the Atlantic Ocean. In the western Atlantic, this species is known from Nova Scotia to Venezuela. They are uncommon in the Gulf of Mexico and Caribbean Sea.

HABITAT AND HABITS: Primarily a coastal pelagic species, Atlantic chub mackerel school by size and occur from the surface to depths of 300 m (985 ft). **OCCURRENCE IN THE CHESAPEAKE BAY:** Atlantic chub mackerel are rare to occasional springtime visitors to the lower Chesapeake Bay.

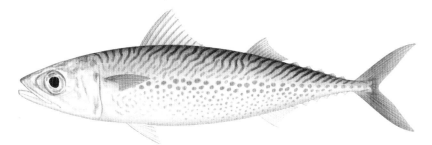

Atlantic chub mackerel *Scomber colias*

REPRODUCTION: A north-south migration brings individuals to the Chesapeake Bay region for overwintering and spawning. There spawning occurs offshore during winter, most often at water temperatures of 15–20°C (59–68°F). The larvae are more common south of Cape Hatteras.

FOOD HABITS: Atlantic chub mackerel are opportunistic feeders on crustaceans, small fishes, and squids. Adults feed at night in the water column and spend the day near the bottom.

IMPORTANCE: In the Chesapeake Bay area, chub mackerel are not sufficiently abundant or predictable to be of commercial or recreational importance.

Atlantic mackerel - *Scomber scombrus* Linnaeus, 1758

KEY FEATURES: Front and back portions of eye covered by adipose eyelid; first dorsal fin with 11–13 slender spines; 5 finlets following dorsal and anal fins. **COLOR:** Bluish black dorsally, with wavy blackish oblique-to-transverse streaks or bars; no blotches or lines on lower sides and belly; bright silvery on lower sides and belly. **SIZE:** Maxi-

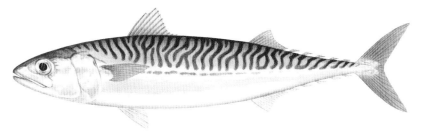

Atlantic mackerel *Scomber scombrus*

mum adult size to about 55 cm (1.8 ft) FL. Females are typically larger than males. RANGE: Both coasts of the northern Atlantic and throughout the Mediterranean. In the western Atlantic, from Labrador to North Carolina.

HABITAT AND HABITS: This species is pelagic, schools by size, and is most abundant in cold and temperate shelf areas. Schools of Atlantic mackerel are dense and can number in the thousands. Their depth range extends to about 180 m (590 ft). They are fast swimmers that overwinter in deeper waters as far south as Cape Hatteras but move north and closer to shore in spring when water temperatures are 11–14°C (52–57°F). OCCURRENCE IN THE CHESAPEAKE BAY: Atlantic mackerel are occasional visitors to the lower Chesapeake Bay in early spring. Most are gone from our area by mid-June to spend their summer and early fall north of Cape Cod. Some will return to the lower Chesapeake Bay in late fall.

REPRODUCTION: In the Chesapeake Bay region, spawning occurs offshore in early spring and progresses to more northerly waters during the summer. Atlantic mackerel do not commence spring spawning until the water temperature reaches at least 8°C (46°F).

FOOD HABITS: Atlantic mackerel are primarily plankton feeders, subsisting on pelagic crustaceans as well as fish eggs and fry.

IMPORTANCE: In the Chesapeake Bay area, this species is of minor importance. In the bay region, most Atlantic mackerel are taken offshore by anglers from mid-February through mid-April. The general low abundance and low variety of fishes available during early spring makes this species popular with anglers who are willing to venture offshore.

King mackerel - *Scomberomorus cavalla* (Cuvier, 1829)

KEY FEATURES: Lateral line abruptly curving downward below second dorsal fin; dorsal fins contiguous; first dorsal fin with 12–17 (usually 15) slender spines. COLOR: Iridescent blue green dorsally; sides plain silver without bars or spots in adults; no black area on anterior third of first dorsal fin; juveniles with small bronze spots in five or six irregular rows. SIZE: Maximum adult size 1.7 m (5.8 ft) FL; females grow larger than males and may live as long as 14 years, with some reports of fish over 20 years of age. RANGE: Massachusetts to Brazil, including the Gulf of Mexico and Caribbean Sea.

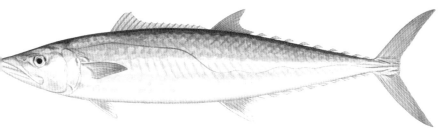

King mackerel *Scomberomorus cavalla*

HABITAT AND HABITS: King mackerel are a surface-dwelling nearshore species that is often found around wrecks, towers, reefs, and other structures. Large schools of similar-sized king mackerel migrate over considerable distances along the Atlantic coast. Some apparently overwinter in the Gulf of Mexico; others may move farther south. OCCURRENCE IN THE CHESAPEAKE BAY: King mackerel are occasional visitors to the lower Chesapeake Bay and rare to occasional in the mid-upper bay. They occur during the warm months of the year (June to October), with peak abundance in September.

REPRODUCTION: Spawning occurs over the middle and outer portions of the continental shelf from May through September along the Atlantic coast.

FOOD HABITS: Fishes are the primary component of king mackerels' diet, but shrimps and squids are also eaten.

IMPORTANCE: King mackerel are an important species for recreational and commercial fisheries throughout their range. In the Chesapeake Bay they are of minimal commercial importance but are occasionally taken in pound nets near the bay mouth. King mackerel are usually common in late summer along the Virginia and North Carolina coasts. Anglers pursue king mackerel by slow trolling, drifting, or anchoring using a variety of live, cut, or artificial baits.

Spanish mackerel - *Scomberomorus maculatus*
(Mitchill, 1815)

KEY FEATURES: Lateral line gradually curving downward to caudal peduncle; dorsal fins contiguous, the first with 17–19 (usually 19) spines. **COLOR:** Metallic greenish to bluish green dorsally, silvery along sides and belly; sides marked with about three rows of round-to-elliptical dusky spots (orange in life) and no longitudinal stripes; anterior third of first dorsal fin black. **SIZE:** Maximum adult size about 77 cm (2.5 ft) FL. **RANGE:** Cape Cod to Florida and coastally from Florida to Yucatán. Florida waters are considered to hold the greatest abundance of Spanish mackerel.

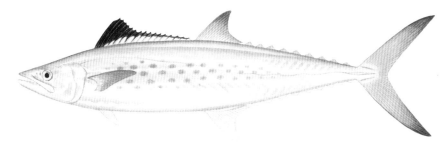

Spanish mackerel *Scomberomorus maculatus*

HABITAT AND HABITS: Spanish mackerel are a surface-dwelling, nearshore species that undertakes long-distance migrations in large schools, traveling along the shore. Schools frequently enter tidal estuaries. **OCCURRENCE IN THE CHESAPEAKE BAY:** Spanish mackerel are common visitors to the Chesapeake Bay from spring to autumn, extending as far north as the mouth of the Sassafras River near the head of the bay. With increasing water temperatures in late February, Spanish mackerel migrate northward and westward from Florida waters, entering the Chesapeake Bay by May, when water temperatures exceed about 17°C (63°F). They spread out in the lower bay, with the major concentration along the lower Western Shore. Spanish mackerel return in autumn to Florida waters, where they overwinter.

REPRODUCTION: Spawning occurs off Virginia from late spring through late summer. Spanish mackerel can attain eight years of age.

FOOD HABITS: Their food consists mainly of small fishes, with lesser amounts of shrimps and squids.

IMPORTANCE: Of commercial and recreational importance throughout their range. In the late 1800s, the Spanish mackerel fishery in the Chesapeake Bay was the largest in the United States. The abundance of Spanish mackerel in the bay has fluctuated greatly since then. The commercial landings in Chesapeake Bay waters over the past 50 years have averaged 54 mt (119,070 lb) per year, with a high of 242.7 mt (535,153 lb) in 1990 and a low of 0.3 mt (662 lb) in 1979. The commercial catch comes primarily from the lower western bay south of the Potomac River and is taken by pound nets, with gill nets and haul seines also making minor contributions. Bay anglers pursue Spanish mackerel by trolling feathers or pork rind or by casting fly and spinning lures into surface schools. In a typical year, about 24,000 Spanish mackerel will be caught in bay waters of Maryland and Virginia, with Virginia waters producing about two-thirds of the catch. The peak fishing months are June through August.

Northern sennet - *Sphyraena borealis* DeKay, 1842

Guaguanche - *Sphyraena guachancho* Cuvier, 1829

KEY FEATURES: Guaguanche have fewer scales in a lateral line series than northern sennet (102–119 vs. 115–130); in guaguanche, the origin of the pelvic fins is slightly more anterior than the origin of the dorsal fin (in northern sennet, the origin of the pelvic fins is more posterior than the origin of the dorsal fin); the appressed pectoral fin reaches to, or extends beyond, the pelvic-fin origin in guaguanche, while in northern sennet, the tip of the appressed pectoral fin does not reach to the pelvic-fin origin; northern sennet have a fleshy tip on the lower jaw, vs. no fleshy tip on the lower jaw in guaguanche. COLOR: Guaguanche have a faint yellow or golden lateral stripe in life, with the top of the head dark. Northern sennet have a dark longitudinal stripe along the side that is sometimes broken into blotches, with the top of the head and the snout black. SIZE: The maximum adult size of both species is about 50 cm (1.6 ft) TL. RANGE: Guaguanche have a broader range than northern sennet. Guaguanche are known from both sides of the Atlantic. In the western Atlantic, guaguanche are known from Massachusetts to Brazil, including the Gulf of Mexico and Caribbean Sea. Northern sennet are known from Nova Scotia to Florida, throughout the Gulf of Mexico, and along the Central American coast to Panama.

HABITAT AND HABITS: Guaguanche are a schooling species that occurs in shallow, generally turbid, coastal waters, over muddy bottoms and often around estuaries.

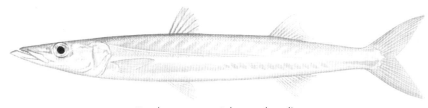

Northern sennet *Sphyraena borealis*

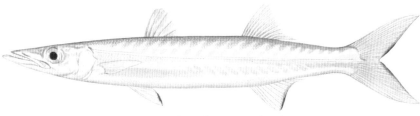

Guaguanche *Sphyraena guachancho*

Northern sennet inhabit coastal waters to depths of 65 m (213 ft) and form large schools over all kinds of substrates, but especially over muddy bottoms. OCCURRENCE IN THE CHESAPEAKE BAY: Northern sennet are more common in the Chesapeake Bay than guaguanche. Whereas guaguanche are rare visitors to the lower bay, juvenile northern sennet may be found most summers near the bay mouth.

REPRODUCTION: Spawning of northern sennet occurs off Florida during winter. Little or no information exists about reproductive habits of guaguanche.

FOOD HABITS: Both species prey mainly upon small fishes and shrimps.

IMPORTANCE: Neither guaguanche nor northern sennet have commercial or recreational value in the Chesapeake Bay region, because of their rarity.

Northern stargazer - *Astroscopus guttatus* Abbott, 1860

KEY FEATURES: Head flattened and about as wide as deep; mouth broad and nearly vertical, with fringed lips; eyes small and located dorsally; nasal openings fringed, posterior openings located behind eyes and formed into elongate grooves; mouth broad; lips fringed; short skin-covered poisonous spine (cleithral spine) located just above pectoral-fin base. COLOR: Upper part of body dusky, with many small irregular white spots that increase in size posteriorly; lower body grayish, with obscure blotches; caudal peduncle with dusky midlateral stripe; first dorsal fin blackish; second dorsal fin with several oblique bars; caudal fin with alternating black and white longitudinal stripes; pectoral fin dusky, with white margin. SIZE: Maximum adult size to about 50 cm (1.6 ft) TL; commonly to about 30 cm (1 ft) TL. RANGE: Found in coastal waters from New York to North Carolina.

Northern stargazer *Astroscopus guttatus*

HABITAT AND HABITS: Northern stargazers are found from inshore waters to depths of 36 m (118 ft). These fish lie buried just below the surface of the sediment with only the lips, the top of the head, and the eyes exposed. The large vertically oriented mouth bears fringes on the lips, presumably to prevent sand from entering the mouth. The nasal openings are also externally fringed, and unlike all but a few fishes, northern stargazers have nasal passages that open into the mouth cavity, thus allowing them to respire without taking water in through the mouth when buried. Stargazers have a pair of specialized organs behind the eyes that can produce an electrical charge. In adults, these organs have been reported to produce as much as 50 volts of electricity, and it is speculated that the discharge is used both to drive off predators and attract prey (the electrical charge is considered too weak to stun prey). OCCURRENCE IN THE CHESAPEAKE BAY: Northern stargazers are year-round residents of the lower Chesapeake Bay and are encountered in the mid-upper bay only during autumn.

REPRODUCTION: Spawning occurs in the lower bay in May and June, but little is known of the early life history of this species.

FOOD HABITS: Northern stargazers are predators that lie in wait for unsuspecting prey to pass near the partially buried mouth. Their diet consists of small fishes and crustaceans.

IMPORTANCE: Of no commercial or recreational value. Occasionally collected in pound nets and trawls near the bay mouth.

Silver perch (Virginia perch, King William perch) - *Bairdiella chrysoura* (Lacepède, 1802)

KEY FEATURES: Mouth moderate and terminal, slightly oblique; preoperculum with serrate margin; scales present on bases of pectoral and pelvic fins and extend onto and cover most of soft dorsal, anal, and caudal fins; dorsal fin continuous but notched nearly to its base, with 10–11 spines in anterior portion and 1 spine and 19–23 soft rays in posterior portion. COLOR: Body greenish or bluish gray above, sides and belly silvery; upper body with faint dusky stripes along scale rows; fins yellowish. SIZE: Maximum size to about 30 cm (1 ft) TL. RANGE: Found from New York to the Gulf coast of northern Mexico.

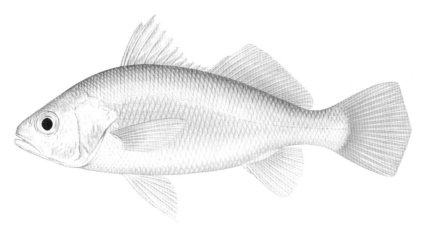

Silver perch *Bairdiella chrysoura*

HABITAT AND HABITS: This inshore species is abundant in shallow coastal and estuarine waters throughout its range. Silver perch occur over both sand and mud bottoms and will occasionally enter freshwater. OCCURRENCE IN THE CHESAPEAKE BAY: Silver perch occur in the Chesapeake Bay throughout the year but are most frequently taken from April through November, with peak abundance in September and October. They retreat to deeper bay waters in the colder months and may migrate to coastal waters during especially cold winters. The abundance of silver perch declines in the northern portion of the bay, and the species is rare north of Baltimore.

REPRODUCTION: Spawning occurs along the bay side and sea side of the Eastern Shore beginning in early- to mid-April and lasting through June. Juveniles settle in shallow seagrass beds, where they are abundant.

FOOD HABITS: The diet consists mostly of small crustaceans, with some polychaete worms and fishes.

IMPORTANCE: Silver perch are taken in pound nets in the lower bay but are of limited commercial value, due to their small size. They are often caught by anglers in bay waters with hook and line but are rarely a targeted species. The recreational catch in Virginia's bay waters typically varies from a few thousand to more than 15,000 in most years.

Spotted seatrout (speckled trout) - *Cynoscion nebulosus* (Cuvier, 1830)

KEY FEATURES: Mouth large, pair of enlarged canine teeth at tip of upper jaw; fins scaleless except for several basal rows on dorsal and anal fins; dorsal fins narrowly separated or continuous but deeply notched; anal fin with 2 spines and 9–12 soft rays. COLOR: Body shades of dusky gray above, with iridescent bluish to pinkish reflections, and silvery below; upper sides and back with numerous scattered round black spots, with spotting continuing onto dorsal and caudal fins; other fins pale yellowish green. SIZE: Maximum size to about 90 cm (3 ft) TL and 7.7 kg (17 lb). RANGE: Found from Cape Cod to Mexico but rare north of Delaware Bay.

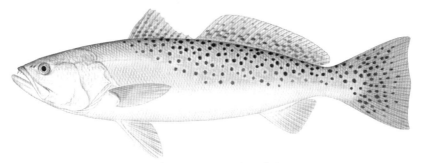

Spotted seatrout *Cynoscion nebulosus*

HABITAT AND HABITS: Spotted seatrout prefer shallow water over sandy bottoms near submerged aquatic vegetation or structures. Adults display a wide salinity tolerance and may be found in salinities as low as 5‰. OCCURRENCE IN THE CHESAPEAKE BAY: Adult spotted seatrout are migratory in the Chesapeake Bay, usually arriving in late April and moving offshore and south in late November. They are most abundant in the lower bay but are found throughout bay waters.

REPRODUCTION: Spawning occurs at night from late May through July near the bay mouth and in nearby coastal waters. In summer and fall, young-of-the-year fish are common in intertidal creeks and may also be found in nearshore beds of submerged aquatic vegetation.

FOOD HABITS: Spotted seatrout are opportunistic carnivores whose food habits change with size; when they are small their diet consists primarily of crustaceans and, as they mature, it shifts to fishes and shrimps. The major food items of adults are shrimps and numerous types of fishes, such as striped mullet, anchovies, and pinfish. Spotted seatrout feed sporadically but primarily during morning hours.

IMPORTANCE: Spotted seatrout are an important component of the recreational fishery in the Chesapeake Bay. In a typical year, anglers in bay waters of Maryland and Virginia will catch more than 60,000 spotted seatrout, with Virginia waters producing 95% of that catch. The largest catches occur from May through November in the lower bay and the York and Rappahannock Rivers. During spring and autumn, spotted seatrout are taken at high tide over shallow eelgrass beds at dawn and dusk, by peeler crab or artificial lures and bait. In November, spotted seatrout are caught by sport fishers in the deep channel areas of the Chesapeake Bay Bridge Tunnel, where they are associated with weakfish. In addition to their importance as a recreational species, spotted seatrout are taken by haul seines, pound nets, and gill nets; however, the catch is of minor commercial importance in the Chesapeake Bay. From 1998 to 2008, the average annual commercial landings in the Chesapeake Bay were 15.2 mt (33,340 lb) per year, with a low of 2.8 mt (6,212 lb) in 2003 and a high of 19.8 mt (43,666 lb) in 2008.

Silver seatrout - *Cynoscion nothus* (Holbrook, 1848)

KEY FEATURES: Mouth large, pair of enlarged canine teeth at tip of upper jaw; snout pointed, with 2 marginal pores; chin without barbels; anal fin with 2 spines and 8–10 soft rays. **COLOR:** Pale grayish brown above, silvery below; back and upper sides sometimes marked with faint rows of spots that follow scale rows; dorsal fin dusky, other fins pale. **SIZE:** Maximum size 38 cm (1.3 ft) TL. **RANGE:** Occurring from the Chesapeake Bay to the Bay of Campeche, Mexico. Silver seatrout are most abundant in the Gulf of Mexico.

HABITAT AND HABITS: Found most often over sandy bottoms in nearshore waters at depths of 10–65 m (33–210 ft). **OCCURRENCE IN THE CHESAPEAKE BAY:** Adult silver

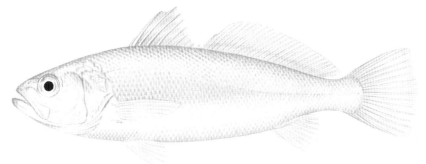

Silver seatrout *Cynoscion nothus*

seatrout are occasionally found in the lower Chesapeake Bay in summer and fall in salinities above 18‰, but they are more common in outside coastal waters. This species is the least abundant of the seatrouts that enter the bay, and little is known of its ecology in Virginia's waters.

REPRODUCTION: Spawning occurs from May through August in nearshore coastal waters in North Carolina. There appears to be no significant spawning in Virginia waters.

FOOD HABITS: Silver seatrout feed on crustaceans and small fishes.

IMPORTANCE: Although silver seatrout are occasionally taken in pound nets near the bay mouth, no commercial fishery exists for this species. Recreational fishers take silver seatrout as a rarity when angling for other species.

Weakfish (gray trout) - *Cynoscion regalis*
(Bloch and Schneider, 1801)

KEY FEATURES: Mouth large, pair of enlarged canine teeth at tip of upper jaw; anal fin with 2 spines and 10–13 soft rays. COLOR: Dark olive green above, with back and sides variously burnished with iridescent blue, coppery, or pinkish reflections; small dusky to blackish blotches and spots forming oblique wavy lines on upper sides; underside white or silvery; anal, pectoral, and pelvic fins yellow. SIZE: Maximum size 91 cm (3 ft) TL and weight 8.5 kg (19 lb). RANGE: Occur from Nova Scotia to about Cape Canaveral, Florida. Weakfish are most abundant from North Carolina through Long Island.

Weakfish *Cynoscion regalis*

HABITAT AND HABITS: Weakfish are typically found over sand or sand/mud bottoms in shallow coastal waters. North of Cape Hatteras, weakfish display a spring and summer migration northward and inshore and a fall and winter movement southward and offshore. OCCURRENCE IN THE CHESAPEAKE BAY: Larger fish (year two and older) appear in the lower bay in April to May, with year-one fish becoming abundant in summer. In the Chesapeake Bay, adult weakfish occur in schools and prefer shallow sandy bottom areas in salinities above 10‰. However, weakfish are found throughout bay waters.

REPRODUCTION: Maturity is reached at age one to two, and spawning takes place near the Chesapeake Bay mouth and in adjacent nearshore waters. The spawning period is protracted (April through August), with peak spawning from May through June. Larvae are taken throughout the lower bay in late summer, and young-of-the-year fish of about 4 cm (1.6 in) TL appear in low-salinity river habitats in August. The young fish grow rapidly in the rivers through October. At about 12 cm (4.7 in) TL they begin to move into more saline waters, and they apparently leave the bay by early winter. North of Cape Hatteras, weakfish tend to be larger after age one and attain a greater longevity than those farther south.

FOOD HABITS: Weakfish feed on a variety of small fishes, larger zooplankton, shrimp, and crabs and become increasingly piscivorous with age.

IMPORTANCE: Weakfish landings by pound nets, gill nets, and haul seines constitute an important fishery in the lower bay, but the fishery has been in decline since at least the late 1980s. Maryland and Virginia waters reported landings of 1,046 mt (2.3 million lb) in 1988, but the annual catch since then has averaged only about half the 1988 amount (544 mt, or 1.2 million lb) and has not exceeded 1,000 mt (2.2 million lb) since that time. More than 80% of the commercial landings come from Virginia's bay waters. Weakfish are a major recreational species in the bay; however, the recreational catch is also showing a major decline in numbers. In 1998, more than 600,000 weakfish were landed in the Chesapeake Bay. In 2002, this number was 266,000, and in 2009, it was only 13,500. Much of the recent decline in the weakfish fishery appears to be attributable to poor natural recruitment.

Spot (Norfolk spot) - *Leiostomus xanthurus*
Lacepède, 1802

KEY FEATURES: Head short, with blunt snout and small inferior mouth; chin without barbels. **COLOR:** Body bluish gray above, brassy white below, with fins pale to yellowish; 12–15 oblique dusky bars on upper sides; distinct dusky to black spot just behind upper end of gill opening. **SIZE:** Maximum size 35 cm (1.2 ft) TL; common to 25 cm (10 in) TL. **RANGE:** Known from the Gulf of Maine to the Bay of Campeche off northern Mexico. Spot are most abundant from the Chesapeake Bay through the Carolinas.

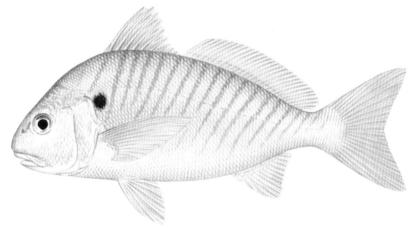

Spot *Leiostomus xanthurus*

HABITAT AND HABITS: Spot are schooling fish that migrate seasonally between coastal and estuarine waters. Adult spot are primarily found in salinities above 5‰, but young-of-the-year often penetrate tributaries to the lower reaches of freshwater. Spot often congregate over oyster beds. **OCCURRENCE IN THE CHESAPEAKE BAY:** Spot are found throughout the Chesapeake Bay. Adults and juveniles enter the Chesapeake Bay during the spring and remain until fall, when they migrate south of Cape Hatteras. Larval spot enter the bay in winter and spring and appear in nursery habitats (low-salinity tidal creeks) in April and May at about 2.5 cm (1 in) TL. The young spot grow rapidly through the summer and by fall have reached an average length of about 12.5 cm (5 in) TL. Most young-of-the-year spot leave the bay by December, but some apparently overwinter.

REPRODUCTION: Spawning occurs at age two to three, in offshore coastal waters in late fall to early spring, with a peak in February. After spawning, adult spot may remain offshore.

FOOD HABITS: Young spot feed predominately on crustacean zooplankton, but they make the transition to bottom-feeding as they grow. Both juveniles and adults are nocturnal predators on polychaete worms, small crustaceans, and mollusks.

IMPORTANCE: Spot are one of the bay's most important commercial and recreational species. Commercial catches come from pound nets, gill nets, and haul seines, with landings in Virginia typically 20 times greater than those in Maryland. Landings of spot in bay waters have shown considerable interannual fluctuation but peaked in 1949 at 3,955 mt (8.7 million lb) and have generally declined since. The last large catch was in 1970, when 2,924 mt (6.4 million lb) were landed. Chesapeake Bay commercial catches in recent years (1998–2008) have averaged 1,590 mt (3.5 million lb) and have only exceeded 2,000 mt (4.4 million lb) 3 times in that period. Recreational anglers take spot from shore or boat while bottom-fishing with a wide variety of baits. Oftentimes, anglers drift-fish over the bottom, using small hooks baited with bloodworms. The spot season in the bay extends from about May through October and peaks in late summer to early fall. The total recreational catch of spot in bay waters typically exceeds 3.2 million fish per year.

Southern kingfish - *Menticirrhus americanus*
(Linnaeus, 1758)

KEY FEATURES: Chin with single short barbel and 2 pairs of lateral pores; dorsal fin continuous but deeply notched; anterior portion with 9–10 (typically 10) flexible spines, posterior portion with 1 spine and 20–26 soft rays; anal fin with 1 spine and 6–8 (typically 7) rays. **COLOR:** Upper body silvery gray or tan, often with with seven or eight faint oblique bars, second and third bars form a faint V shape below spinous dorsal fin; may be almost uniformly colored; belly silvery white; spinous dorsal fin dusky along margin; soft dorsal fin pale; caudal fin dusky, upper portion often with tan pigment, lower portion dusky; pectoral fin dusky to black; pelvic and anal fins white to yellowish. **SIZE:** Maximum size about 42 cm (1.4 ft) TL and 1 kg (2.5 lb). A common size is about 28 cm (11 in) TL and less than 0.5 kg (1 lb) in weight. **RANGE:** Occurring from New York south through the Gulf of Mexico to Buenos Aires, Argentina. Southern kingfish are most common from the Chesapeake Bay to the Bay of Campeche, Mexico, but are rare to absent in South Florida and the Antilles.

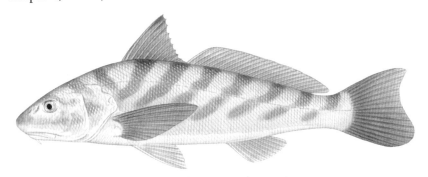

Southern kingfish *Menticirrhus americanus*

HABITAT AND HABITS: Southern kingfish are demersal and occur over a wide variety of substrates from mud to sand-mud mixtures; adults prefer sand bottoms of ocean beaches and mouths of large coastal sounds. Southern kingfish are found in a wider range of water temperatures (8–30°C, or 46–86°F) and salinities (6–35‰) than either gulf kingfish or northern kingfish. **OCCURRENCE IN THE CHESAPEAKE BAY:** Southern kingfish enter the Chesapeake Bay in the spring and leave in the fall for wintering grounds on the continental shelf. They are most abundant in the polyhaline waters of the lower bay.

REPRODUCTION: Spawning occurs outside bay waters from April through September. This species is estuarine-dependent and uses estuaries as nurseries. Juveniles grow quickly in the estuaries, at a rate of almost 25 mm (1 in) per month. This species may attain five to six years of age.

FOOD HABITS: Southern kingfish are bottom feeders, primarily on small crustaceans and polychaete worms. They are active feeders at night, using touch and smell to find their prey.

IMPORTANCE: See discussion under northern kingfish. In some years southern king-fish are more abundant in the bay than northern kingfish. For instance, in the years 2003–2005 more than 110,000 southern kingfish were caught each year by anglers in Virginia's bay waters, while the catch of northern kingfish during this same period averaged fewer than 70,000 fish per year.

Gulf kingfish - *Menticirrhus littoralis* (Holbrook, 1847)

KEY FEATURES: Chin with single short barbel; dorsal fin continuous but deeply notched; anterior portion with 10–11 (typically 10) flexible spines, posterior portion with 1 spine and 19–26 soft rays; anal fin with 1 spine and 6–8 (typically 7) rays. COLOR: Upper body silvery gray, with bronze cast to sides and on cheeks; belly white; nape and sides without bars or stripes; spinous dorsal fin light brown, with dusky tip; caudal fin pale, with dusky margin and blackish tip on upper lobe; other fins pale to dusky. SIZE: Maximum size about 46 cm (1.5 ft) TL. RANGE: Occur from Delaware south and throughout the Gulf of Mexico to Brazil. This species is most common south of Cape Hatteras and in the Gulf of Mexico.

HABITAT AND HABITS: As with other members of this genus, they are more frequent-ly found along coastal beaches, especially in small schools within the surf zone. Gulf

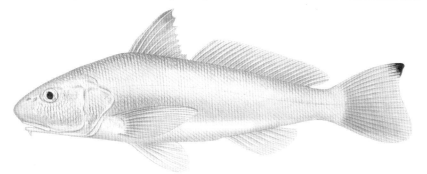

Gulf kingfish *Menticirrhus littoralis*

kingfish will also inhabit shallow offshore waters with sand or mud bottoms. This species is less tolerant of lower salinity waters than other kingfish. OCCURRENCE IN THE CHESAPEAKE BAY: Gulf kingfish are the least common of the three species of *Menticirrhus* that occur in the bay and are distributed within the Chesapeake Bay primarily in the higher saline waters near the bay mouth.

REPRODUCTION: As with other kingfish, gulf kingfish spawn from late spring to early autumn.

FOOD HABITS: This species is a bottom feeder that preys on crustaceans, worms, and shellfish.

IMPORTANCE: See discussion under northern kingfish.

Northern kingfish - *Menticirrhus saxatilis*
(Bloch and Schneider, 1801)

KEY FEATURES: Chin with single short barbel; dorsal fin continuous but deeply notched; anterior portion with 10–11 flexible spines, the second or third elongate; anal fin with 1 spine and 7–9 (typically 8) rays. COLOR: Upper body silvery gray or tan, with multiple dusky oblique bars, five or six of them prominent; bar that precedes and bar that follows spinous dorsal fin converge above pectoral-fin base to form prominent V shape; dusky longitudinal streak on posterior flank that extends onto caudal fin in larger specimens; margin of caudal and pectoral fins dusky, other fins

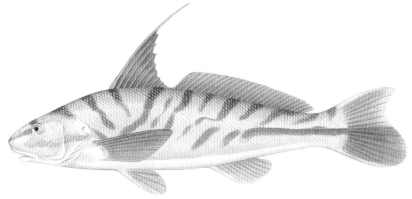

Northern kingfish *Menticirrhus saxatilis*

pale to dusky. SIZE: Maximum size about 50 cm (1.8 ft) TL and 1.3 kg (3 lb). RANGE: Found from Maine to the northern coast of Yucatán. Northern kingfish are most common from New York through North Carolina.

HABITAT AND HABITS: These fish are common in the surf zone and in estuaries, frequenting shallow waters over sand or sandy mud bottoms. Juveniles will enter tidal zones of rivers. OCCURRENCE IN THE CHESAPEAKE BAY: Northern kingfish enter the Chesapeake Bay in April and May and leave in the fall for wintering grounds on the continental shelf. They are most abundant in the lower bay in salinities above 10‰ over firm bottoms. This species has been collected as far north in the bay as the Chester River.

REPRODUCTION: Spawning occurs at ages two to three from May through August in coastal waters. Young-of-the-year appear on coastal and estuarine beaches in July, with recruitment occurring through the summer and early fall. The growth rate of this species is rapid, 1.8–2.4 mm (0.07–0.09 in) per day, one of the fastest growth rates for fishes in the region.

FOOD HABITS: This species is a bottom feeder, primarily on small crustaceans, other fishes, and polychaete worms.

IMPORTANCE: This species, also called sea mullet, roundhead, or whiting, is of limited commercial importance and is taken along with other species by gill nets, haul seines, and pound nets. Catch data, as well as anglers, rarely distinguish this species from southern kingfish (*Menticirrhus americanus*) and gulf kingfish (*M. littoralis*), with which it is often coincidental. Commercial catches of kingfish in the bay have been insignificant since 1950. Recreational anglers take kingfish while bottom-fishing over firm bottoms using a variety of baits. The fishing season runs from April through mid-November but is best from August through October. Based on recreational catch statistics from 1998–2009, the annual catch of northern kingfish is quite variable, ranging from a high of 96,000 fish in 2004 to a low of 6,000 fish in 2005. On average during this 12-year period, approximately 32,000 northern kingfish were caught each year by anglers in the Chesapeake Bay region. The kingfish is considered a very tasty food fish.

Atlantic croaker (hardhead) - *Micropogonias undulatus* (Linnaeus, 1766)

KEY FEATURES: Mouth horizontal, inferior; chin with three to five pairs of small barbels and five pores; preopercular margin strongly serrate. COLOR: Upper sides with numerous brassy spots, forming oblique wavy bars; bars less distinct in large individuals; dorsal fin with numerous dusky spots that form indistinct dusky streaks in posterior portion; anal and pelvic fins pale to yellowish. SIZE: Maximum size usually no bigger than 50 cm (1.8 ft) TL. Croakers north of Cape Hatteras typically attain a larger average size than those south; the largest ever recorded was 69 cm (2.3 ft) TL, from New Point Comfort Lighthouse near the York River mouth. RANGE: In the western Atlantic from Massachusetts to Florida and throughout the Gulf of Mexico. This species is uncommon north of New Jersey.

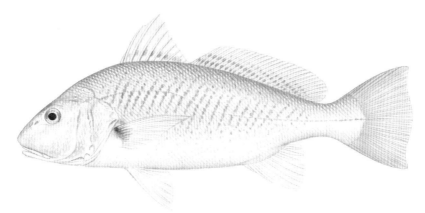

Atlantic croaker *Micropogonias undulatus*

HABITAT AND HABITS: A bottom dweller that frequents muddy or sandy areas, Atlantic croakers are one of the most abundant inshore fishes along the southeastern coast of the United States. Found to depths as great as 100 m (328 ft) but also very common in estuaries. OCCURRRENCE IN THE CHESAPEAKE BAY: Adults move into the Chesapeake Bay in April and are found throughout the bay, most often in salinities above 5‰. This species has been collected as far north in the bay as the Susquehanna flats. Young-of-the-year croakers of about 20 mm (0.8 in) TL enter the estuary beginning in August and move into the nursery habitat of low-salinity to freshwater creeks. In autumn the young croaker move into the deeper portions of tidal rivers, where they overwinter and leave the bay with adults the following fall. Atlantic croaker display greater interannual variability in abundance than any other bay fish. Evidence suggests that these fluctuations may be weather related, with colder winters causing increased mortality in overwintering young-of-the-year.

REPRODUCTION: The first spawning occurs at age two to three in continental shelf waters from July through February, with peak spawning from August through October.

FOOD HABITS: Atlantic croakers are opportunistic bottom feeders consuming polychaete worms, mollusks, a variety of small crustaceans, and occasionally small fishes.

IMPORTANCE: Atlantic croakers form the basis of an important but highly variable commercial fishery in the Chesapeake Bay. Atlantic croakers are taken with pound nets, gill nets, and haul seines, with Virginia waters providing about 90% of the landings. Commercial catches from the early 1960s through 1974 were usually well below 750 mt (1.6 million lb) but increased to almost 2,500 mt (5.5 million lb) in 1975. Another downward trend began in 1980, with another big rebound in 1993. Since 1993, the catch has averaged 5,148 mt (11.3 million lb) per year. Croakers are a mainstay of the recreational fishery. Between 1998 and 2009, almost 8 million croakers were caught each year in Maryland and Virginia bay waters. Croakers are taken by anglers from mid-April through September while bottom-fishing with a variety of baits. The bay (and world) record is a 3.9 kg (8.7 lb) fish caught near New Point Comfort Lighthouse (Mobjack Bay near the York River mouth) in 2007.

Black drum - *Pogonias cromis* (Linnaeus, 1766)

KEY FEATURES: Lower jaw with 5 pores and 10–13 pairs of barbels. **COLOR:** Body blackish, with brassy luster above; grayish white below; fins dusky to black; young with four to six vertical bars on sides. **SIZE:** Maximum adult size 1.7 m (5.6 ft) TL and 66 kg (146 lb). Black drum are the largest member of the family Sciaenidae on the Atlantic coast. **RANGE:** Coastal waters and estuaries from Massachusetts to Argentina. Black drum are uncommon north of Delaware Bay.

HABITAT AND HABITS: Typically found in coastal areas with sand or soft bottoms. Frequently found near clam or oyster beds. **OCCURRENCE IN THE CHESAPEAKE BAY:** Adult black drum enter the Chesapeake Bay in April and concentrate just north of the bay mouth west of Cape Charles, where spawning occurs from mid-to-late

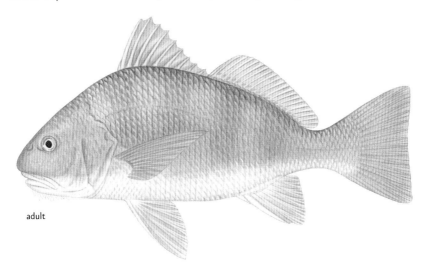

adult

Black drum *Pogonias cromis*

juvenile

April and continues through early June. After the end of spawning, adult black drum spread out in the bay; they are recorded as far north in the bay as the Elk River near the head of the bay. Black drum migrate southward in late fall.

REPRODUCTION: When vernal warming commences, black drum spawn in bays, estuaries, or inlets. Males attract females with drumming noises emanating from the swim bladder.

FOOD HABITS: Black drum use their sensory chin barbels to detect bottom-dwelling prey and are able to crush clams, oysters, mussels, and crabs by means of their pharyngeal tooth plates.

IMPORTANCE: The commercial catch of black drum in Virginia is limited to 54.4 mt (120,000 lb), while in Maryland the commercial catch and sale of black drum is banned in the Chesapeake Bay and its tributaries. A small commercial gillnet fishery exists on the sea side of the lower Delmarva Peninsula in the early spring and on the bay side late in the season. The catch shows great annual fluctuation, with recent catches varying from 14.0 mt (30,835 lb) in 2002 to 51.6 mt (113,858 lb) in 2003. An intense four-to-six-week recreational fishery in late April to early June is centered on and around the spawning ground west of Cape Charles. The best time to catch black drum is during a full moon. Anglers frequently use a hook baited with soft crab or clam and drift-fish on the bottom in 5–6 m (16–20 ft) of water. During the period 1998–2009, Maryland and Virginia anglers caught an average of 8,235 black drum per year, with more than 90% of that catch coming from Virginia waters.

Red drum (channel bass) - *Sciaenops ocellatus*
(Linnaeus, 1766)

KEY FEATURES: Mouth inferior, horizontal; chin without barbels. COLOR: Color somewhat variable, but most frequently silvery tinged with pale to dark copper or brassy above, white below; dorsal and caudal fins dusky, anal and pelvic fins white, tinged with yellow; pectoral fins rusty at tips; one to many black ocellated spots on upper portion of caudal peduncle, spots may also be present on caudal-fin base and upper body posteriorly. SIZE: Maximum size about 1.3 m (5 ft) TL and 42 kg (92 lb). RANGE: Known from the Gulf of Maine to the northern coast of Mexico but uncommon north of New Jersey. Red drum are more abundant in the Gulf of Mexico than along the Atlantic coast.

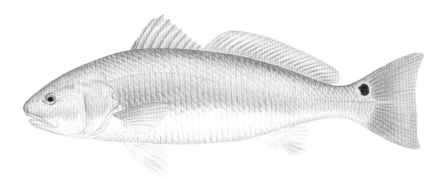

Red drum *Sciaenops ocellatus*

HABITAT AND HABITS: Adult red drum are most common in nearshore marine waters, where they may travel in large schools. Red drum occasionally move offshore, but there is no specific information available on the timing, duration, or extent of these movements. Red drum display a northerly migration in the spring and southerly movement in the fall. OCCURRENCE IN THE CHESAPEAKE BAY: Adult red drum occur in the Chesapeake Bay from May through November and are most abundant in the spring and fall near the bay mouth. Red drum may occasionally overwinter in the bay during mild winters. This species extends as far north in the bay as the Patuxent River.

REPRODUCTION: Spawning occurs in nearshore coastal waters from late summer through fall, with young-of-the-year appearing in estuaries from August through September. Red drum can live to age 50 or older.

FOOD HABITS: The food of red drum consists of small- to moderate-size crustaceans and fishes.

IMPORTANCE: Red drum do not constitute an important fishery in the Chesapeake Bay area. Virginia's commercial catch, as high as 83 mt (180,000 lb) in 1950, has been insignificant since then, averaging less than 6 mt (13,216 lb) per year. Maryland's annual catch has not exceeded 0.5 mt (1,103 lb) since 1994. A modest recreational fishery exists, with most fish taken by surfcasting from seaside beaches and some by bait fishing along the bay side of the lower Eastern Shore. The largest catch in recent years was in 2009, when almost 57,000 red drum were landed from Virginia's bay waters.

Star drum - *Stellifer lanceolatus* (Holbrook, 1855)
Banded drum - *Larimus fasciatus* Holbrook, 1855

KEY FEATURES: The body of star drum is slightly longer than that of banded drum. In star drum, the head is broad, with greatly enlarged sensory canals, giving the top of the head a spongy appearance; banded drum lack this feature. The preopercular margin of star drum is serrated, with four to six distinct spines; the preopercular margin of banded drum is smooth. For both species, the middle rays of the caudal fin are longest; however, in star drum, the caudal fin is pointed. **COLOR:** In banded drum, there are seven to nine dusky bars on the back extending to the body midline. Star drum lack bars. **SIZE:** The maximum adult size of both species is about 20 cm (8 in) TL, with banded drum sometimes attaining 23 cm (9 in) TL. **RANGE:** Star drum are found in inshore waters from Virginia to Texas but are not common north of South

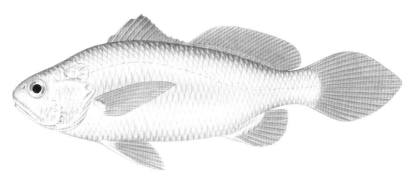

Star drum *Stellifer lanceolatus*

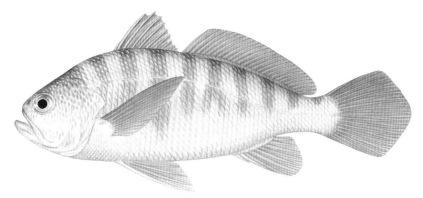

Banded drum *Larimus fasciatus*

Carolina. Banded drum range from Massachusetts to Florida and the northern Gulf of Mexico but are not abundant north of Cape Hatteras.

HABITAT AND HABITS: Star drum are typically found over sand-mud bottoms in coastal waters to depths of about 20 m (66 ft). Star drum are also encountered at the mouths of tidal rivers. Banded drum are only rarely encountered in estuaries and are more typically found over mud and sandy mud bottoms in coastal waters to depths of about 60 m (200 ft). OCCURRENCE IN THE CHESAPEAKE BAY: Star drum sometimes stray northward during the summer and enter the Chesapeake Bay mouth, whereas banded drum are occasionally found in salinities above 15‰ in the lower Chesapeake Bay.

REPRODUCTION: Spawning of star drum occurs in early summer in nearshore and estuarine waters south of the Chesapeake Bay. With banded drum, spawning occurs offshore from late spring to early autumn, with a peak during the summer months.

FOOD HABITS: Both star drum and banded drum feed in the water column, and both species feed predominately on small crustaceans.

IMPORTANCE: Because of their small size and low abundance, neither species is commercially or recreationally important. Banded drum are occasionally caught in pound nets, and anglers sometimes take banded drum on hook and line in the lower bay.

Sand drum - *Umbrina coroides* Cuvier, 1830

KEY FEATURES: Snout rounded, projecting beyond mouth; mouth inferior; chin with short stout barbel. **COLOR:** Body silvery, back and upper sides with nine black bars and undulating longitudinal streaks; belly and lower sides pale. **SIZE:** Maximum adult size about 35 cm (1.1 ft) TL. **RANGE:** Found coastally from the Chesapeake Bay to Brazil, including much of the Gulf of Mexico and Caribbean Sea. Sand drum are absent from the northern Gulf of Mexico and much of the Central American coast, and they are uncommon north of Florida.

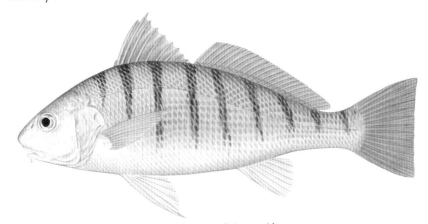

Sand drum *Umbrina coroides*

HABITAT AND HABITS: Sand drum are typically found in shallow waters along sandy beaches. These fish can also be encountered over muddy bottoms in estuaries and sometimes near coral reefs. They are most common over sand bottoms but can be found over mud. **OCCURRENCE IN THE CHESAPEAKE BAY:** Sand drum are rare visitors to the lower Chesapeake Bay in late summer.

REPRODUCTION: No information on the reproductive habits of sand drum is available.

FOOD HABITS: Sand drum feed on bottom-dwelling organisms.

IMPORTANCE: None in the Chesapeake Bay. A commercial fishery for sand drum exists in more tropical waters, where the species is more abundant.

Atlantic spadefish - *Chaetodipterus faber*
(Broussonet, 1782)

KEY FEATURES: Body very deep and compressed; head blunt; mouth small and terminal; spinous and soft fins separate; spinous portion of dorsal fin low in adults, however, juveniles have prominent third dorsal-fin spine, 9 dorsal-fin spines with 21–23 soft rays; anterior portion of soft dorsal and anal fins prolonged into filaments. COLOR: Four to six dusky vertical bars on silvery gray background, bars fading with age; first bar passes through eye, last bar on caudal peduncle; most of fins blackish. Juvenile Atlantic spadefish typically all black. SIZE: Maximum adult size has been reported as 90 cm (3 ft) TL. RANGE: Known from New England to southern Brazil, including the Gulf of Mexico and Caribbean Sea, but rare north of the Chesapeake Bay. This is the only family member native to the western Atlantic Ocean.

HABITAT AND HABITS: Atlantic spadefish are coastal fishes inhabiting depths to about 30 m (98 ft) or greater. Most Atlantic spadefishes frequent rocky bottoms and reefs

Atlantic spadefish *Chaetodipterus faber*

as well as wrecks and pilings, sometimes forming dense schools. OCCURRENCE IN THE CHESAPEAKE BAY: Atlantic spadefish are occasional to common visitors during the summer and autumn to the mid-lower Chesapeake Bay, extending as far north as the mouth of the Patuxent River. They occur in schools from a few to more than 500 individuals, most frequently around towers, buoys, or other structures.

REPRODUCTION: Spawning occurs offshore from May to August, and large spawning aggregations can be found near the surface on warm sunny days when the water temperature is 24–29°C (75–85°F). Young-of-the-year are common in nearshore habitats of the lower bay in the late summer and early fall. Atlantic spadefish attain at least 8 years of age and more than 9 kg (20 lb) in weight.

FOOD HABITS: Atlantic spadefish feed on the bottom or near the surface on a wide variety of invertebrates, including jellyfishes, hydroids, polychaetes, amphipods, sponges, and sea anemones.

IMPORTANCE: Atlantic spadefish are of only minor commercial importance, but they are pursued by anglers fishing near the Chesapeake Bay mouth and offshore. Virginia's bay waters yield about 20,000 spadefish per year. Because of their small mouth, small hooks baited with pieces of jellyfish, clam, or mussels are used. Most catches run 1.5–3.5 kg (about 3–8 lb); however, the Chesapeake Bay record for Atlantic spadefish is 6.4 kg (14 lb). The flesh is of excellent quality, but this fish has been associated with ciguatera poisoning, which is caused by toxins in the flesh of certain tropical marine fishes.

Bluefish - *Pomatomus saltatrix* (Linnaeus, 1766)

KEY FEATURES: First dorsal fin low, with 7 or 8 slender spines; second dorsal fin high, with 1 spine and 23–28 soft rays; anal fin with 3 spines (1 or 2 may be small and difficult to see) and 23–27 soft rays; caudal fin forked. COLOR: Greenish blue along the back, silvery below; black blotch present at base of pectoral fin. SIZE: Maximum adult size is about 1.1 m (3.6 ft) TL; however, typical bluefish caught in Chesapeake Bay waters are about 30 cm (11.8 in) TL. RANGE: Occur in temperate and subtropical waters around much of the world except the eastern Pacific. In the western Atlantic, bluefish occur from Nova Scotia to Brazil, including the Gulf of Mexico, but they are rare or absent in most of the Caribbean Sea.

HABITAT AND HABITS: Bluefish are found along coastlines and in estuaries. This species is a migratory pelagic fish that primarily travels in schools. These schools are

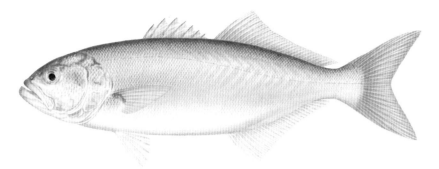

Bluefish *Pomatomus saltatrix*

generally groups of like-sized fish that can form large aggregations. OCCURRENCE IN THE CHESAPEAKE BAY: Bluefish, visitors to Chesapeake Bay waters from spring to autumn, are abundant in the lower bay and common most years in the mid-upper bay, although they are rare north of Baltimore. In early autumn, bluefish begin to migrate out of the bay and move south along the coast. Peak abundances near the bay mouth occur in April to July and again in October to November.

REPRODUCTION: In the Chesapeake Bay area, peak spawning is in July over the outer continental shelf. Bluefish reach sexual maturity at age 2, when about 36 cm (1.2 ft) TL, and they can live more than 12 years.

FOOD HABITS: Bluefish are voracious predators with the rare reputation of wantonly killing prey they do not eat. They are sight feeders throughout the water column, with smaller individuals feeding on a wide variety of fishes and invertebrates and large bluefish feeding almost exclusively on fishes.

IMPORTANCE: Bluefish are one of the most important sport fishes in the Chesapeake Bay and usually rank in the top five in both weight and number caught every year. Within the Chesapeake Bay, the bulk of the commercial landings of bluefish come from gillnets and pound nets. The peak commercial catch in recent years occurred in 1998, when just under 445 mt (980,000 lb) were landed, of which more than 80% was from Virginia waters. Bluefish abundance displays considerable year-to-year variation and also long-term cycles. These fluctuations in abundance probably have multiple causes, some of which are likely to be natural factors. However, in recent years overfishing has also become a concern. Bluefish are well known to anglers for their incredible biting power and tendency to strike at almost any object. Recreational landings of bluefish in bay waters of Maryland and Virginia average more than 450,000 fish each year.

Skilletfish - *Gobiesox strumosus* Cope, 1870

KEY FEATURES: Head broadly rounded and flattened, body compressed posteriorly; opercle with sharp spine; pelvic fins united and modified into large sucking disk. **COLOR:** Dusky to pale gray background coloration, pattern variable; some individuals with pale crossbars. **SIZE:** Maximum adult size 8 cm (3 in) TL; common to 5–6 cm (2–2.5 in) TL. **RANGE:** In the western Atlantic from New Jersey to Florida, throughout the northern Gulf of Mexico, and southward along the Caribbean coast of South America to southeastern Brazil. In the Eastern Pacific from the Gulf of California to Ecuador.

HABITAT AND HABITS: Skilletfish are typically small inconspicuous benthic fishes with a ventral sucking disk formed from highly modified pelvic fins. The sucking disk enables these fish to cling to rocks, shells, or other objects. Skilletfish are usually found in shallow water or intertidal zones with grass or rocks, often occurring in waters less than 1 m (3.3 ft) in depth, but they can be found as deep as 33 m (110 ft). **OCCURRENCE IN THE CHESAPEAKE BAY:** Skilletfish are year-round residents of the Chesapeake Bay that are common to abundant throughout most of the bay, extending as far north as the Susquehanna River mouth. Skilletfish are typically found on oyster bars and in eelgrass beds. During the warmer months skilletfish are found closer to shore, and in the winter they retreat to deeper water. Skilletfish are tolerant of the high temperatures and low-oxygen conditions that occur seasonally in the Chesapeake Bay; they are able to decrease their metabolic rate when environmental conditions are stressful.

REPRODUCTION: Spawning occurs in empty oyster shells from April through August. The male guards the eggs until hatching.

FOOD HABITS: Skilletfish feed primarily on small crustaceans such as amphipods and isopods, as well as polychaete worms.

IMPORTANCE: Of no commercial or recreational interest, but occasionally collected in seines. Skilletfish readily adapt to aquariums, where they typically adhere to the glass.

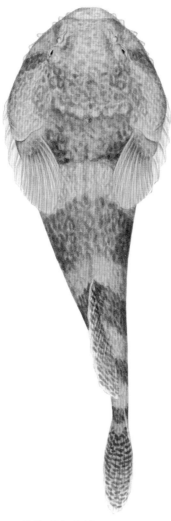

Skilletfish *Gobiesox strumosus*

Lyre goby - *Evorthodus lyricus* (Girard, 1858)
Darter goby - *Ctenogobius boleosoma*
(Jordan and Gilbert, 1882)

KEY FEATURES: Lyre gobies are somewhat stouter than darter gobies. In lyre gobies, the teeth are bicuspid in females and males smaller than 60 mm (2 in) SL, while the teeth are conical with pointed tips in darter gobies. The caudal fin of lyre gobies is long, with a somewhat rounded margin in females, and pointed in large males. The caudal fin of darter gobies is pointed and arrow-shaped, and longer in males. COLOR: Lyre gobies have irregular lateral body markings, but usually with some indication of five or six vertical bars and faint median blotches; the first dorsal fin of males has several ocelli and is marked with narrow brown lines in females and young males; the caudal-fin base has distinctive upper and lower dusky blotches, and large males have one or two ocelli on the upper part of the caudal fin. Darter gobies are predominantly pale tan, with large males being occasionally dusky; the sides have four to five narrow longitudinal brown spots or bars along the midline, with those under the second dorsal fin diverging upward to form a distinct "V"; a large dusky spot above the pectoral-fin base; the dorsal and caudal fins are streaked with small brown spots and the caudal fin of older males sometimes has two pink stripes. SIZE: Maximum adult size of lyre gobies is about 77 mm (3 in) TL, while darter gobies are slightly smaller, with a maximum adult size of about 62 mm (2.5 in) TL. RANGE: Lyre gobies range from the Chesapeake Bay to Surinam, including the Gulf of Mexico, while darter gobies are known from Delaware Bay to Brazil, including the Gulf of Mexico. Darter gobies are rarely encountered north of Cape Hatteras.

HABITAT AND HABITS: Lyre gobies inhabit fresh and brackish waters in shallow muddy estuarine environments, as well as tidal marshes and ponds, and are not common anywhere along the Atlantic coast. Darter gobies frequently inhabit mud and sand bottoms, in freshwater to saltwater in lower estuaries and sounds, and they also occur on shell bottoms. OCCURRENCE IN THE CHESAPEAKE BAY: Both lyre gobies and darter gobies are rarely encountered in the Chesapeake Bay. Lyre gobies have most frequently been recorded from the Lynnhaven River sub-estuary near the bay mouth.

REPRODUCTION: Spawning of darter gobies occurs in coastal waters from March to August. Little is known of the spawning habits of lyre gobies.

FOOD HABITS: Darter gobies feed by taking mouthfuls of sediment and sifting out the

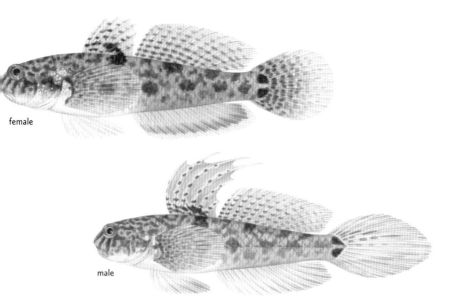

female

male

Lyre goby *Evorthodus lyricus*

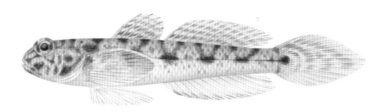

Darter goby *Ctenogobius boleosoma*

small interstitial organisms with their gill rakers; copepods and ostracods comprise the primary food items. Lyre gobies are detritus feeders.

IMPORTANCE: Neither lyre gobies nor darter gobies have commercial or recreational value.

Naked goby - *Gobiosoma bosc* (Lacepède, 1800)

KEY FEATURES: Body lacking scales; first dorsal fin with 7 flexible spines; second dorsal fin typically with 1 spine and 12 rays; total anal-fin elements 11. **COLOR:** Body brownish green along back and sides, pale on belly; nape and sides with eight to ten narrow pale crossbars; males more dusky than females, especially when guarding territory. **SIZE:** Maximum adult size to about 60 mm (2.4 in) TL. **RANGE:** Known from Massachusetts to Cape Canaveral, Florida, and along the northern Gulf of Mexico from Pearl Bay, Florida, to Campeche, Mexico.

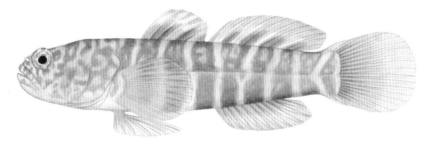

Naked goby *Gobiosoma bosc*

HABITAT AND HABITS: Naked gobies inhabit protected bays and estuaries throughout their range. They are often associated with oyster reefs. Naked gobies can tolerate low- to moderate-salinity conditions and have been collected in freshwater. **OCCURRENCE IN THE CHESAPEAKE BAY:** Naked gobies are year-round residents of the Chesapeake Bay that are common to abundant throughout most of the bay, extending as far north as the mouth of the Susquehanna River. Naked gobies inhabit fresh to marine shallow waters (as shallow as 10 cm, or 4 in) and are common on vegetated flats, on oyster reefs, and among the growth on pilings, seawalls, and other firm substrates. Naked gobies may retreat to channels or bury themselves in muddy bottoms in the winter.

REPRODUCTION: Spawning begins in late spring, when water temperatures exceed 20°C (68°F), and extends to early fall. The most common nesting sites are in dead shells. After hatching, the larvae are planktonic until they reach about 12–15 mm (0.5–0.6 in) TL. Naked gobies are the most abundant larval species in ichthyoplankton surveys in low-salinity waters of the Chesapeake Bay. At about 20 days after hatching, planktonic larvae have been observed to school just above oyster reefs prior to settling.

FOOD HABITS: Naked gobies feed on small crustaceans and worms.

IMPORTANCE: Naked gobies are of no direct commercial or recreational value. However, based on the density of larvae, the population of this species in the Chesapeake Bay must be exceedingly large and must therefore constitute a major forage resource for other fishes.

Seaboard goby - *Gobiosoma ginsburgi*
Hildebrand and Schroeder, 1928

KEY FEATURES: Body entirely scaleless except for pair of large ctenoid scales on each side of caudal-fin base; first dorsal fin with 7 flexible spines; second dorsal fin typically with 1 spine and 11 rays. **COLOR:** Body brownish, with six or seven whitish crossbars; lateral line with longitudinally elongate dusky spots; males darker, especially when guarding territory. **SIZE:** Maximum adult size about 52 mm (2 in) TL. **RANGE:** Known from Massachusetts to Georgia. The original description of this species was based on specimens collected from several localities in the Chesapeake Bay.

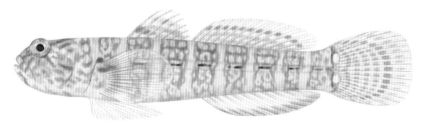

Seaboard goby *Gobiosoma ginsburgi*

HABITAT AND HABITS: As with other gobies, seaboard gobies are secretive and cryptic. Seaboard gobies occur in shallow water estuaries to depths of about 30 m (98 ft) on the continental shelf. They are most common on hard bottoms of shell or rubble. Within estuaries, they are most abundant in meso- to polyhaline waters of at least 2 m (6.6 ft) in depth. **OCCURRENCE IN THE CHESAPEAKE BAY:** Seaboard gobies are year-round residents of the Chesapeake Bay that are common to abundant in the lower bay and occasional to common in the upper bay. This species typically inhabits deeper flats and oyster reefs from spring to autumn, retreating to channels in winter. These are the most abundant gobies in open waters of the bay.

REPRODUCTION: Spawning occurs from May through October. Dead shells are the primary spawning sites. Females lay a small mass of eggs, each attached by an adhesive stalk to the underside of dead shells or other firm overhanging substrate. The eggs are guarded and tended by the male. Spawning may also occur offshore.

FOOD HABITS: Seaboard gobies feed on a variety of invertebrates, primarily small crustaceans.

IMPORTANCE: Seaboard gobies are of no direct commercial or recreational value; however, they constitute an important forage resource for other fishes.

Green goby - *Microgobius thalassinus*
(Jordan and Gilbert, 1883)

KEY FEATURES: Tongue bilobate; body mostly with scales, about 45–50 in lateral series; first dorsal fin with 7 flexible spines; second dorsal fin typically with 1 spine and 15 rays; total anal-fin elements 16; caudal fin nearly lanceolate. COLOR: Males have three to five golden-tan bars behind pectoral fin, the first extending to shoulder above pectoral-fin base; first dorsal fin reddish; anal fin has submarginal row of dark spots. Females have reddish spot near front of first dorsal fin and several submarginal black spots near back of fin; anal fin lacks submarginal row of dark spots of males. SIZE: Maximum adult size about 40 mm (1.6 in) SL. RANGE: Known from the Chesapeake Bay to Galveston, Texas, but apparently absent from southeast Florida and the Florida Keys.

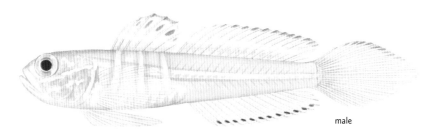

male

Green goby *Microgobius thalassinus*

HABITAT AND HABITS: Green gobies frequent shallow waters with soft bottoms; they apparently prefer mud or muddy sand bottoms. These fish may be burrowers, as several other species of this genus are known to dig burrows in soft substrate. Most collections of green gobies have been made in depths of less than 6 m (20 ft). This species can tolerate a wide range of salinities, from near freshwater to full seawater.

OCCURRENCE IN THE CHESAPEAKE BAY: Green gobies are year-round residents of the Chesapeake Bay that are occasional to common in the lower reaches of tributaries throughout the bay. They are frequently found in mud and oyster habitats in the bay, often in association with the red beard sponge. In winter months, green gobies retreat to channels or channel edges.

REPRODUCTION: Green gobies spawn from June through September. Although little is known about the spawning habits of this species, it is assumed that they follow the pattern of other Chesapeake Bay gobies. Green gobies apparently are short-lived, with a lifespan of only a year or so.

FOOD HABITS: The food of green gobies consists of small crustaceans, including amphipods.

IMPORTANCE: Of no direct commercial or recreational value but serve as a forage resource for other fishes.

White perch - *Morone americana* (Gmelin, 1789)

KEY FEATURES: Dorsal-fin spines large and strong, with 7 to 11 spines; second dorsal fin with 1 spine followed by 10–13 soft rays; anal fin with 3 strong spines and 8–10 soft rays; second anal-fin spine approximately same length as third, but stouter. **COLOR:** Color variable, mostly silvery, often greenish to bluish and blackish on back and sometimes brassy on sides; dorsal and caudal fins dusky. **SIZE:** Maximum adult

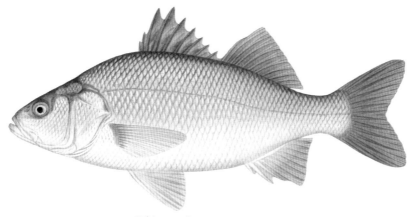

White perch *Morone americana*

size is about 48 cm (1.6 ft) TL. RANGE: Known from Nova Scotia to South Carolina but most abundant from the Hudson River to the Chesapeake Bay.

HABITAT AND HABITS: White perch are ubiquitous in estuaries and freshwater eco-systems, living in waters ranging in salinity from zero to full-strength-seawater, but they usually inhabit waters less than 18‰. Because white perch tolerate a wide range of salinities, they become easily acclimated in freshwater ponds and other impound-ments. White perch frequent areas with level bottoms composed of compact silt, mud, sand, or clay and show little preference for vegetation, structures, or other shelter. OCCURRENCE IN THE CHESAPEAKE BAY: White perch are abundant year-round residents found in all tributaries of the Chesapeake Bay. From spring through autumn white perch are present on flats and in channels, and they retreat to deep channels in winter.

REPRODUCTION: White perch exhibit semianadromous spawning migrations and move into the fresh- to low-salinity waters of large rivers, where they spawn from April through June when water temperatures are about 11–16°C (52–61°F). The young utilize the quiet-water shore margins of the spawning areas as nursery grounds.

FOOD HABITS: The diet of white perch changes with age and habitat. Juveniles feed on aquatic insects and small crustaceans, while larger white perch prey on crabs, shrimps, and small fishes.

IMPORTANCE: White perch are one of the most important recreational and commer-cial fishes in the Chesapeake Bay, especially in Maryland waters, where most of the bay landings of white perch occur. Commercial landings of white perch in the bay peaked in 1969 at about 976 mt (2.2 million lb) and have generally declined since. Chesapeake Bay landings of white perch from 1999–2008 averaged about 640 mt (1.4 million lb) per year, with Maryland waters providing more than 90% of that to-tal. Commercial landings are made with a variety of gear types, especially haul seines, fyke nets, pound nets, and gillnets. Catches are greatest during the spring spawning season and also from September through November, when white perch school to feed on migrating herring. The recreational fishery for white perch is significant, especially in Maryland. In 2007, recreational landings in Maryland waters were esti-mated to exceed 2.9 million white perch, which is the highest total in recent years. The recreational fishery is concentrated in the spring and autumn, when white perch are taken by drifting live bait or by trolling artificial lures near the surface.

Striped bass (rockfish) - *Morone saxatilis*
(Walbaum, 1792)

KEY FEATURES: Dorsal fins separate; anal fin with 7–13 (typically 11) soft rays; second anal-fin spine shorter than third. **COLOR:** Olive green, blue, or black along back with sides silvery, becoming white on belly; sides with seven or eight narrow dusky stripes, which follow scale rows; stripes frequently interrupted and obscure on young fish. **SIZE:** Maximum adult size approaches 1.8 m (6 ft) TL. Females grow larger and live longer than males. Fish that are 0.94 m (3.1 ft) weigh approximately 9.5 kg (21 lbs) and are about 11 years old; this is about the maximum size and age for male striped bass. Striped bass older than 11 years are typically females, with ages exceeding 30 years reported. **RANGE:** The natural range of striped bass extends along the Atlantic coast from the St. Lawrence River in Canada to the St. Johns River in Florida and in the Gulf of Mexico from western Florida to Louisiana.

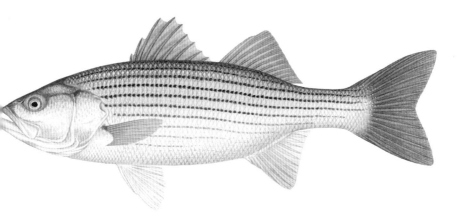

Striped bass *Morone saxatilis*

HABITAT AND HABITS: Striped bass are tolerant of a variety of environmental conditions and may be found in marine, estuarine, or riverine habitats, depending on the season. They are anadromous, moving from the ocean or other high-saline waters to tidal freshwaters to spawn. **OCCURRENCE IN THE CHESAPEAKE BAY:** Striped bass are abundant year-round residents found in all tributaries of the bay. During summer and winter, striped bass are found in deep bay channels. In autumn, they are more concentrated in the lower reaches of rivers.

REPRODUCTION: Chesapeake Bay tributaries constitute the principal spawning areas for striped bass along the mid-Atlantic coast, and spawning activity is most intense in the first 40 km (25 mi) of freshwater over sand or mud bottom. Spawning migrations begin as early as March, with peak spawning activity at the end of April or early May when water temperatures are 13–20°C (55–68°F). Spawning behavior often involves a single large female being courted simultaneously by a number of smaller males. Larval nurseries are the nearshore areas of the spawning sites and the brackish waters immediately downstream. Larvae are often found in association with white perch larvae.

FOOD HABITS: Adult and juvenile striped bass are predators on a variety of fishes, crustaceans, squids, mussels, and worms; larval striped bass eat small planktonic crustaceans.

IMPORTANCE: Striped bass are one of the most important recreational and commercial species in the Chesapeake Bay. Commercial landings of striped bass from the Chesapeake Bay averaged 1,936 mt (4.3 million lb) per year from 1999 to 2008, which approximates the levels from the 1950s and 1960s and is a reversal of the drastic decline of striped bass stock in the 1980s. Increased regulation, along with estuarine habitat protection and restocking efforts, are likely responsible for this reversal. Commercial landings are made primarily with haul seines, gillnets, and pound nets, whereas recreational fishermen troll artificial lures or bottom-fish with natural baits such as eel. Casting for breaking fish is also an effective method. The recreational catch of striped bass in Maryland and Virginia waters in 2006 exceeded 1 million individuals, the highest total in recent years. Recent cases of an infectious disease, mycobacteriosis, in Chesapeake Bay striped bass have raised concern about the impact of this disease on the overall health of the coastal striped bass population.

Tessellated darter - *Etheostoma olmstedi* Storer, 1842

KEY FEATURES: First dorsal fin arched, with 8–10 spines; second dorsal fin slightly separate from first dorsal fin and slightly higher, with 10–17 soft rays; caudal fin truncate to slightly rounded. COLOR: Pale olive on back, shading to whitish on sides and belly; sides and back with eight or more dusky irregular X- or W-shaped markings; dorsal, caudal, and pectoral fins spotted and barred with black; males with black blotch between first two spines of first dorsal fin; pelvic and anal fins black during breeding season. SIZE: Maximum adult size about 9 cm (3.5 in) TL; however, adults are typically in the 4–6 cm (1.8–2.4 in) TL range. RANGE: St. Lawrence Seaway west to Lake Ontario and southward along the Atlantic Coast to the St. Johns River drainage in Florida.

Tessellated darter *Etheostoma olmstedi*

HABITAT AND HABITS: This species is a bottom dweller frequently found on sandy substrates in clear slow-moving freshwater streams. They can also be found on mud bottoms in quiet standing waters. Tessellated darters are known to bury themselves in the substrate so that only their eyes are visible. **OCCURRENCE IN THE CHESAPEAKE BAY:** Tessellated darters are common to abundant in all tributaries of the Chesapeake Bay and have been collected in brackish waters with salinities as great as 13‰.

REPRODUCTION: As with other darters, male tessellated darters assume breeding colors during spawning. This results in the fins and body of the male becoming duskier, almost to the point of being black. The male also develops fleshy pads at the tips of the pelvic-fin rays during spawning season. Spawning occurs from late April to June, with the eggs being deposited on the underside and sides of rocks, logs, or other debris. The eggs are guarded and cared for by the male. Tessellated darters may exceed three years of age.

FOOD HABITS: Tessellated darters feed primarily on bottom-dwelling insects such as midge larvae but will also eat small crustaceans, snails, and algae.

IMPORTANCE: Due to their small size, tessellated darters are of no recreational or commercial value except as food for larger species.

Yellow perch (yellow ned) - *Perca flavescens*
(Mitchill, 1814)

KEY FEATURES: Dorsal fins well separated; 11–15 spines in first dorsal fin; 1–3 spines and 12–16 soft rays in second dorsal fin; caudal fin slightly forked. COLOR: Dusky olive green on back and yellow on sides; back and sides with six to eight black or brown crossbars; anal, pelvic, and pectoral fins red or orange, brightest in males during spawning season. SIZE: Maximum adult size is about 45 cm (1.5 ft) TL. RANGE: Yellow perch are widely distributed in the upper Mississippi Valley, many eastern states, the Great Lakes, and Canada. Along the East Coast, yellow perch are known from Canada to South Carolina.

Yellow perch *Perca flavescens*

HABITAT AND HABITS: Yellow perch often travel in schools and inhabit cool-water lakes and reservoirs as well as coastal rivers, streams, and low-salinity estuaries. They are frequently associated with shoreline areas where moderate amounts of vegetation provide food, cover, and spawning habitat. Adults can be found in moderate currents but prefer sluggish currents or slack-water habitat, particularly during spawning. OCCURRENCE IN THE CHESAPEAKE BAY: Yellow perch are common to abundant in most tributaries of the Chesapeake Bay and are sometimes found in brackish water (<13‰) at river mouths.

REPRODUCTION: Yellow perch require freshwater for spawning and begin spawning migrations from river mouths into tributaries in late February and early March. Spawning occurs at night in shallow water when water temperatures range from 7–13°C (45–55°F). Eggs are laid in gelatinous strands in and around aquatic vegetation and submerged tree branches; some exposed strands can be observed at low tide.

FOOD HABITS: The diet of yellow perch is diverse and includes insect larvae, crustaceans, and small fishes. They are known to be cannibalistic.

IMPORTANCE: Yellow perch are of commercial and recreational importance in the Chesapeake Bay because of the flavor of their flesh and their proclivity to take a baited hook. Most are caught during their spring spawning runs. Commercial yellow perch landings primarily come from fyke nets, with pound nets, gillnets, and haul seines contributing lesser amounts. Maryland typically provides more than 80% of the commercial landings from bay waters. As with other anadromous fishes in the region, declines in abundance were noted in the 1970s and 1980s. In Maryland and Virginia waters, commercial landings of yellow perch averaged 60 mt (132,000 lb) per year from 1950 to 1969 but only 23 mt (50,600 lb) per year from 1970 to 1989. From 1990 to 2008, the average annual commercial landings almost doubled from the 1970–1989 period, to 43 mt (94,700 lbs). Maryland recently banned commercial fishing of yellow perch until March 14th to allow them to migrate to their spawning grounds. These fish are of excellent eating quality and are popular with shore anglers, who use a variety of natural baits.

Striped blenny - *Chasmodes bosquianus* (Lacepède, 1800)

KEY FEATURES: Flap-like structure (cirrus) above eye small to absent in adults, decreasing in size with increasing SL; dorsal fin very long, continuous, and joined to caudal fin; 10–12 dorsal-fin spines, typically 11, total dorsal-fin elements typically number 29–30; caudal fin rounded. **COLOR:** Varies from pattern of pale longitudinal lines on olivaceous background to pale blotches or diffuse bands; head with small dark spots dorsally and laterally; dusky spot at base of caudal fin; males greenish, with series of eight or nine irregular bluish lines that converge near the tail and yellowish-orange band that courses through dorsal fin, with bright blue spot near front of fin; caudal fin of males reddish orange, as is posterior part of dorsal fin; anal fin of males dusky, with white membranous tips; females are darker olive green, with pale green longitudinal lines and series of dusky bars more distinct near tail. **SIZE:** Maximum adult size 8 cm (3 in) SL. **RANGE:** Known from New York southward to northern Florida. Rarely encountered north of the Chesapeake Bay.

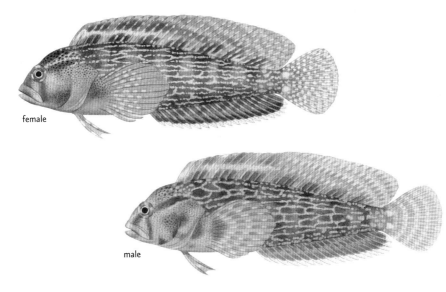

female

male

Striped blenny *Chasmodes bosquianus*

HABITAT AND HABITS: Oyster bars are the typical habitat of striped blennies, but they can be found in grassy areas as well as on rubble or sand bottoms. This species usually inhabits waters that range in salinity from 15–25‰. **OCCURRENCE IN THE CHESAPEAKE BAY:** Striped blennies are common to abundant residents of the entire Chesapeake Bay and are usually found in grass beds or over sand bottoms or hard substrates such as oysters reefs. Striped blennies inhabit shallow areas during spring and summer and move to deeper flats and reefs in autumn and to channels during winter.

REPRODUCTION: The spawning season begins in mid-March and extends through August. Dead shells are the preferred spawning sites, and males vigorously defend their established spawning territory.

FOOD HABITS: The food of these blennies is primarily small crustaceans and mollusks.

IMPORTANCE: Of no direct commercial or recreational value.

Feather blenny - *Hypsoblennius hentz* (Lesueur, 1825)

KEY FEATURES: "Feathery" cirrus above eye long and branched; dorsal fin long and continuous, attached to caudal fin, 11–13 dorsal-fin spines, typically 12; total dorsal-fin elements typically number 25–27; caudal fin rounded; anal fin lower than dorsal fin and not attached to caudal fin. COLOR: Head and body with multitude of small deep-brown spots, varying in intensity and sometimes forming irregular bands; spots more numerous on head and larger posteriorly on body; large dusky ovoid area posterior to eye; underside of head with two or three chevron-shaped bands; dorsal fin of males typically with bright blue spot overlapping second spine, spot pale in females; pelvic fins nearly black; anal fin of males dark, with pale margin, barred in females; other fins variously barred or spotted. SIZE: Maximum adult size 10 cm (4 in) SL. RANGE: Known from Nova Scotia southward to Florida, around the Florida tip, and westward to the Yucatán Peninsula.

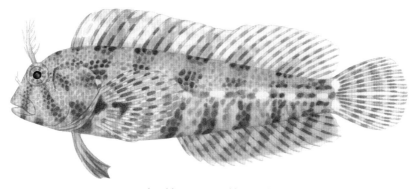

Feather blenny *Hypsoblennius hentz*

HABITAT AND HABITS: Feather blennies typically inhabit oyster reefs and rocky shores but are also found on mud and grass flats. This species is most often found in waters that range in salinity from 12–30‰. OCCURRENCE IN THE CHESAPEAKE BAY: Feather blennies, year-round residents of the Chesapeake Bay, are common throughout the bay, frequenting grass flats and oyster reefs and firm live-bottom habitats during the summer and deeper channels during the winter.

REPRODUCTION: Spawning extends from May to August; empty oyster shells in live oyster reefs near the low-tide line are the preferred nest sites.

FOOD HABITS: These blennies feed on small crustaceans, mollusks, and sea squirts.

IMPORTANCE: Of no direct commercial or recreational value.

Northern snakehead - *Channa argus* (Cantor, 1842)

KEY FEATURES: Body torpedo-shaped, compressed posteriorly; head long and pointed, flat on top; mouth large, with sharp teeth; dorsal fins continuous, with 51–52 rays; caudal fin rounded; anal fin long, with 33–35 rays; pelvic fins are close together and originate behind pectoral-fin base. COLOR: In fish greater than 10 cm (about 4 in) TL, head and body with dark brown blotches; blotches on head tend to be longitudinal, while body blotches are more vertical, numbering about 12; blotches extend onto dorsal and anal fins; blotches much less apparent in fish smaller than 10 cm (about 4 in) TL. SIZE: Maximum adult size is about 85 cm (2.8 ft) TL. RANGE: The native range of northern snakeheads is East Asia from China to Russia, but they have been introduced elsewhere in Asia and in eastern Europe. As these fish are popular in the aquarium trade and in certain ethnic restaurants, live fish have been shipped around the world. Introductions occurred in a number of U.S. waterways, primarily in the East, and northern snakeheads became established in some.

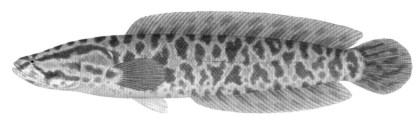

Northern snakehead *Channa argus*

HABITAT AND HABITS: Northern snakeheads are most often found in stagnant shallow ponds, swamps, and slow streams with abundant aquatic vegetation. This species apparently has a limited salinity tolerance (<10‰) but a broad temperature tolerance and is able to live in waters ranging from 0°C to greater than 30°C. They are also capable of surviving in water with very low oxygen content. OCCURRENCE IN THE CHESAPEAKE BAY: Since the discovery of breeding populations in tributaries of the Potomac River near Washington, D.C., in 2004, northern snakeheads are becoming increasingly abundant and moving downriver. A 1.8-kg (4-lb) northern snakehead was caught in a pound net in St. Mary's Creek near the mouth of the Potomac River in August 2010. The significant increase in range of northern snakeheads within the Potomac River has been both rapid and remarkable and demonstrates the adaptability of this species. In 2011 a large northern snakehead was captured in the lower Rappahannock River.

REPRODUCTION: Northern snakeheads build spawning nests in aquatic flora, where females lay eggs that are externally fertilized by males. In Potomac River tributaries, they are repeat spawners between May and September.

FOOD HABITS: The diet of northern snakeheads is diverse and includes insect larvae, amphibians, crustaceans, and other fishes. In Potomac River tributaries, northern snakeheads are largely piscivorous and known to eat banded killifish, bluegill, pumpkinseed, and white perch.

IMPORTANCE: This species is a valuable food fish in its native range, where it is targeted for aquaculture. In the Potomac River system, they have been considered a highly undesirable introduction whose impact on the ecosystem is unknown at present. However, they are worthy game fish, and because they have become firmly established, the Virginia Department of Game and Inland Fisheries has begun a tagging program to learn more about snakeheads' ecology and distribution. Given the latitudinal distribution of their native range (25°N–50°N) northern snakeheads could, in theory, establish populations in many waterways of North America. Import and interstate transport of northern snakeheads were banned in 2002 as the threat posed by them became better understood.

Spotfin butterflyfish - *Chaetodon ocellatus* Bloch, 1787

KEY FEATURES: Head short and deep; snout pointed; teeth arranged in brush-like bands in jaws; dorsal fin continuous, with 12–13 strong spines and 18–21 soft rays; caudal fin gently rounded; anal fin similar to soft portion of dorsal fin, with 3 strong spines and 15–17 soft rays. **COLOR:** Head and body grayish to yellowish, caudal peduncle yellowish orange; jet-black band extending from dorsal-fin origin through eye and across cheek; second, indefinite bar courses from middle of anal fin to middle of soft dorsal fin in juveniles; narrow yellow bar courses along upper margin of gill opening to pectoral-fin base; large blackish blotch near center of soft dorsal-fin base (not evident in all fish) with intense black spot at margin; in small juveniles, basal blotch on dorsal fin often obscured by second body bar; all fins slightly yellowish orange. **SIZE:** Maximum adult size about 18–20 cm (7–8 in) TL. **RANGE:** Inhabit coral reefs from Florida to Brazil. Juveniles are carried northward by the Gulf Stream, accounting for their occurrence as far north as Nova Scotia.

juvenile

Spotfin butterflyfish *Chaetodon ocellatus*

HABITAT AND HABITS: Adults are found in tropical waters on or near coral reefs to depths of about 30 m (98 ft). **OCCURRENCE IN THE CHESAPEAKE BAY:** Spotfin butterflyfish are rare to occasional visitors to the lower Chesapeake Bay during the summer and autumn. All spotfin butterflyfish that enter the bay are juveniles and may grow to a length of 7–8 cm (3 in) TL before succumbing to the low water temperatures of winter. In the bay, they are most common around eelgrass beds.

REPRODUCTION: Other than that pair bonds are formed for spawning, little is known of the reproductive habits of this species.

FOOD HABITS: With their elongate snout, spotfin butterflyfish are able to retrieve food from places inaccessible to other fishes. This species is typically diurnal in habit, feeding from the bottom or on plankton. Prey consists of small invertebrates, such as coral polyps and copepods, although some fish browse on algae. Juveniles are reported to pick parasites off other fishes.

IMPORTANCE: Spotfin butterflyfish are of no commercial or recreational importance. However, they make attractive aquarium fish and will subsist on frozen tubifex worms and brine shrimp. They may be taken in the lower bay with seine nets around shallow grass beds in late summer.

Irish pompano - *Diapterus auratus* Ranzani, 1842

Spotfin mojarra - *Eucinostomus argenteus*
Baird and Girard, 1855

Silver jenny - *Eucinostomus gula* (Quoy and Gaimard, 1824)

KEY FEATURES: Irish pompano differ from spotfin mojarras and silver jennies in having a very deep body and a serrated preopercle. Spotfin mojarras and silver jennies are less deep-bodied than Irish pompano and have the lower edge of the preopercle entirely smooth. Spotfin mojarras are differentiated from silver jennies in that the scaleless patch between the eyes has a narrow opening anteriorly so it is not completely enclosed by scales. The scaleless patch on the snout between the eyes of silver jennies is completely enclosed by scales. COLOR: All three mojarras with silvery body. In Irish pompano, anal and pelvic fins yellowish; in juveniles sides of body typically have four to six pale vertical bars. Spinous dorsal fin in spotfin mojarras often dusky and often has distinct concentration of pigment near fin tip. Spinous dorsal fin in silver jennies has moderate concentration of pigment near tip. SIZE: Irish pompano are larger than either spotfin mojarras or silver jennies and attain a maximum adult size of about 34 cm (1.1 ft) TL. Maximum adult size of spotfin mojarras is about 19 cm (7.5 in) TL and that of silver jennies, about 15 cm (6 in) TL. RANGE: All three are found in coastal waters of the western Atlantic from New Jersey to Brazil, including the Gulf of Mexico and Caribbean Sea.

HABITAT AND HABITS: Irish pompano inhabit shallow coastal waters, where they are common in seagrass beds, mangrove-lined creeks, and lagoons and are often found in low-salinity areas of coastal rivers. Similarly, silver jennies inhabit shallow coastal waters and are abundant in various estuarine habitats, including seagrass beds and mangrove forests, but typically do not penetrate far up coastal rivers. Silver jennies are occasionally taken on the continental shelf. Spotfin mojarra juveniles occur sporadically in estuarine habitats, and adults are often caught over the continental shelf. OCCURRENCE IN THE CHESAPEAKE BAY: These mojarras are rare visitors to the lower Chesapeake Bay during summer. Spotfin mojarras are recorded from only as far north in the bay as the lower Eastern Shore and the York River mouth on the Western Shore. Silver jennies are known from as far north in the bay as the York River.

REPRODUCTION: Little is known about the spawning habits of these species.

FOOD HABITS: With their tube-like protractile mouths, these mojarras feed on bottom-living invertebrates and plant material.

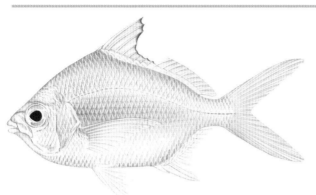

Irish pompano *Diapterus auratus*

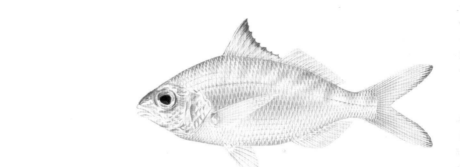

Spotfin mojarra *Eucinostomus argenteus*

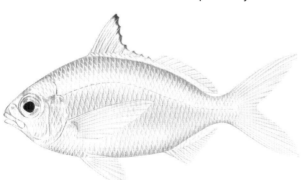

Silver jenny *Eucinostomus gula*

IMPORTANCE: None of the three is of any commercial or recreational value in the bay region.

Sheepshead - *Archosargus probatocephalus*
(Walbaum, 1792)

KEY FEATURES: Front of both jaws with eight broad incisor-like teeth, their edges straight or only slightly notched; three rows of molars laterally in upper jaw, two in lower jaw; procumbent spine in front of dorsal fin. **COLOR:** Body silvery to greenish yellow and marked by incomplete dark bar across top of head, followed by six distinct dark bars on body; fins dusky. **SIZE:** Maximum adult size approaches 90 cm (3 ft) TL; common to 35 cm (1.2 ft) TL. **RANGE:** Three populations of sheepshead are known from Nova Scotia to Rio de Janeiro, Brazil, including the Gulf of Mexico and Caribbean Sea. The population that ranges from Nova Scotia to Cedar Key, Florida, is regarded as a subspecies, *Archosargus probatocephalus probatocephalus*.

Sheepshead   *Archosargus probatocephalus*

HABITAT AND HABITS: Sheepshead frequent jetties, wharves, pilings, shipwrecks, and other structures that become encrusted with barnacles, mussels, and oysters. Sheepshead will enter brackish waters and occasionally freshwater and are not usually found in waters of less than 15.5°C (60°F). **OCCURRENCE IN THE CHESAPEAKE BAY:** Sheepshead are regular summer visitors to the mid-lower Chesapeake Bay, extending as far north as the Potomac River on the Western Shore and the Nanticoke River on the Eastern Shore.

REPRODUCTION: Spawning in U.S. waters occurs offshore during late winter and early spring. After spawning, adults return to nearshore waters, with juveniles inhabiting grass flats. As the fish mature, they disperse to more high-relief, hard-bottom areas. Sheepshead can live longer than eight years.

FOOD HABITS: Sheepshead prey on barnacles, shellfish, and crabs that are crushed by their strong molars and incisors. They are known to scrape bivalves from rocks and pilings.

IMPORTANCE: Where abundant (from North Carolina southward), sheepshead are an important recreational and commercial species. However, in the Chesapeake Bay region, sheepshead are of little or no commercial interest even though they are occasionally taken in pound nets. For the period 1999–2008, the average annual landings of sheepshead were 3 mt (6,700 lb), almost all of which came from Virginia waters. Although not common enough in the bay to be targeted by anglers, they are highly prized when caught. A few thousand sheepshead are usually taken each year in Virginia waters of the bay. Anglers often use live fiddler crab bait when bottom-fishing around structures for sheepshead. They are regarded as excellent food fish.

Pinfish - *Lagodon rhomboides* (Linnaeus, 1766)
Spottail pinfish - *Diplodus holbrookii* (Bean, 1878)

KEY FEATURES: Spottail pinfish are more deep bodied than pinfish and lack a procumbent spine in front of the dorsal fin, which pinfish have. As these fishes are similar in many respects, it is easiest to distinguish them by color. COLOR: Spottail pinfish are brassy on the back, silvery on the sides and belly, with a conspicuous dusky saddle across the caudal peduncle reaching well below the lateral line. Pinfish have a silvery body with yellow and blue stripes and a blackish blotch near the origin of the lateral line. SIZE: Spottail pinfish reach a slightly larger maximum adult size, 46 cm (1.5 ft) TL, than pinfish (40 cm, or 1.3 ft, TL). RANGE: Spottail pinfish range from New Jersey to northern Florida on the Atlantic Coast and from the Florida Keys to southern Texas in the Gulf of Mexico. Spottail pinfish are not common north of Cape Hatteras. Pinfish are known from Cape Cod through the Gulf of Mexico to the Yucatán Peninsula.

HABITAT AND HABITS: Both species are most often found around pilings and jetties and over vegetated bottoms. Although typically a shallow-water species, spottail pinfish also range offshore. Pinfish are tolerant of a wide range of salinities and are sometimes found in brackish waters and freshwater. OCCURRENCE IN THE CHESAPEAKE BAY: Spottail pinfish are rare visitors to the lower Chesapeake Bay during the summer, extending as far north as the York River on the Western Shore and the lower Eastern Shore. Pinfish are occasional visitors to the mid-lower Chesapeake Bay from spring to autumn, extending as far north as the Choptank River.

REPRODUCTION: For both species, spawning occurs offshore from mid-October to March.

FOOD HABITS: Both of these species are daytime feeders. Juveniles are omnivorous, feeding on epiphytes and small invertebrates picked from the substrate. Fishes larger than about 12 cm (4.7 in) SL appear to be mainly herbivorous, feeding on epiphytes and seagrasses.

IMPORTANCE: Neither of these pinfishes is of any commercial interest in the Chesapeake Bay region, and both are caught only incidentally by anglers. Pinfish are caught around docks and pilings, whereas spottail pinfish are caught over hard bottoms. In Virginia waters, catches of pinfish averaged only about 1,100 fish per year from 1998 to 2007. In areas where pinfish are more abundant, they are popular with anglers as live bait.

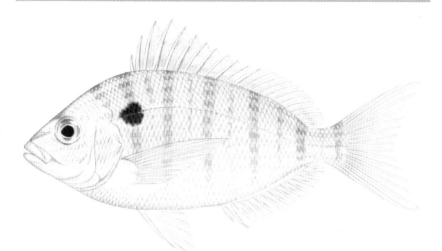

Pinfish *Lagodon rhomboides*

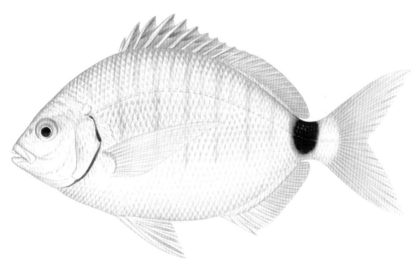

Spottail pinfish *Diplodus holbrookii*

Scup - *Stenotomus chrysops* (Linnaeus, 1766)

KEY FEATURES: Snout pointed; anterior teeth narrow, in close-set bands, not notched, almost conical; two rows of molars laterally in jaws; scales extending onto caudal-fin base and forming low sheath on dorsal and anal fins; no procumbent spine in front of dorsal fin, dorsal fin long and continuous with 12 spines and 12 soft rays, spinous portion high, soft portion rounded. **COLOR:** Bluish silver dorsally, silvery ventrally; blue blotch above eye; side with 12–15 indistinct longitudinal stripes and 6 or 7 faint dusky bars; small black spot dorsally in pectoral-fin axil; dorsal, anal, and caudal fins with blue flecks; blue band on back near dorsal-fin base. **SIZE:** Maximum adult size 45 cm (1.5 ft) TL; common to 25 cm (10 in) TL. **RANGE:** Nova Scotia to eastern Florida, but rare south of North Carolina.

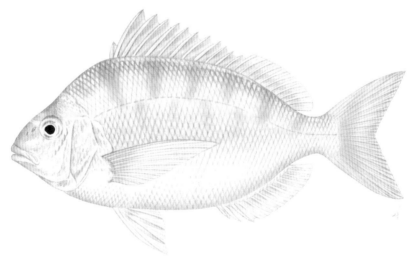

Scup *Stenotomus chrysops*

HABITAT AND HABITS: Scup live in schools over hard-bottom areas and submerged structures at depths of 2–37 m (7–121 ft), where they browse on bottom-dwelling invertebrates. Scup move inshore to coastal areas in spring, where they stay until autumn. Water temperature appears to influence the movement of scup, as they are most frequently found at temperatures of 13–25°C (55–77°F). **OCCURRENCE IN THE CHESAPEAKE BAY:** Scup are common to abundant visitors to the lower Chesapeake Bay from spring to autumn, extending as far north as the York River, and they migrate offshore to deeper waters during winter. Young-of-the-year also inhabit polyhaline waters of the bay from June to October.

REPRODUCTION: Spawning apparently occurs from May through August in nearshore waters north of the Chesapeake Bay; however, eggs and larvae are rarely collected.

FOOD HABITS: With their strong molars, scup are able to crush and consume crabs, sea urchins, snails, and clams.

IMPORTANCE: Scup are of minor commercial and recreational importance in the Chesapeake Bay. Commercial landings in the bay since 1999 have averaged about 70 mt (154,000 lb) per year. Most of the commercial catch of scup is landed by pound nets and haul seines in the fall as the fish begin to move out to deeper water. The recreational catch of scup averages only 2,300 fish per year from Virginia waters. Anglers catch scup by bottom-fishing with natural baits and jigging with small artificial lures. Scup are considered tasty pan-fish.

Gray snapper - *Lutjanus griseus* (Linnaeus, 1758)

KEY FEATURES: Chin pores absent; outer pair of canine teeth in upper jaw much larger than those in lower jaw; tooth patch on roof of mouth V-shaped or crescentic, with posterior extension; dorsal fin continuous, with 10 spines and 14 soft rays (occasionally 13); caudal fin emarginate; anal-fin margin rounded, with 3 spines and 7 or 8 soft rays. COLOR: Variable, depending on surroundings; typically gray, greenish gray, or dusky olive on back and sides, sometimes with reddish tinge, grayish on belly; young specimens with dusky stripe from snout through eye to upper opercle and blue stripe on cheek below eye; margin of spinous dorsal fin dusky in young. SIZE: Maximum adult size about 60 cm (2 ft) TL; common to 45 cm (1.5 ft) TL. RANGE:

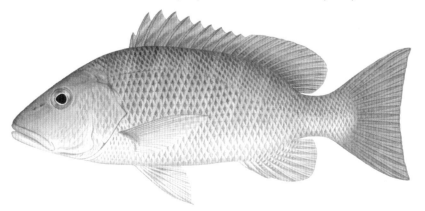

Gray snapper *Lutjanus griseus*

Known from Massachusetts to Brazil, including the Gulf of Mexico and Caribbean Sea. Adults are rare north of the Carolinas.

HABITAT AND HABITS: Gray snapper inhabit coastal as well as offshore waters to depths of 180 m (590 ft), with larger fish generally found offshore. Inshore, gray snapper are found over smooth bottoms and around pilings, rock piles, seagrass meadows, and mangrove thickets. Offshore, this species frequents irregular bottom areas such as coral reefs, shipwrecks, and rocky areas. Gray snapper can tolerate a wide range of salinities and are sometimes found in freshwater. OCCURRENCE IN THE CHESAPEAKE BAY: Small juvenile gray snapper, typically about 2.5 cm (1 in) TL on arrival, are regular visitors to the lower Chesapeake Bay during the summer and fall, extending as far north as the Rappahannock River. The largest specimen recorded from the Chesapeake Bay was 9.2 cm (3.6 in) TL.

REPRODUCTION: The breeding population does not extend north of Cape Hatteras, but to the south, spawning occurs offshore from April to November, with a peak during the summer. Spawning occurs in aggregations around the time of the full moon.

FOOD HABITS: Gray snapper often form large aggregations in the daytime and disperse at night as they move into grass flats to feed. The diet changes with age and consists mainly of a variety of small fishes and crustaceans. Gray snapper become more piscivorous with increasing size.

IMPORTANCE: In the Chesapeake Bay region, gray snapper are of no commercial or recreational importance.

Pigfish - *Orthopristis chrysoptera* (Linnaeus, 1766)

KEY FEATURES: Chin with 2 anterior pores and central groove just behind symphysis of lower jaw; dorsal fin continuous, rather low, its origin over or slightly in advance of pectoral-fin base; dorsal-fin with 12 or 13 spines followed by 15 or 16 rays, spines rather slender, sharp; caudal fin deeply concave, upper lobe longest; anal fin with 3 spines and 12 or 13 rays; pectoral fin longer than pelvic fin. COLOR: Bluish gray along back, shading to silver on sides and belly; each body scale with blue center, edge with bronze spot, these spots forming very distinct orange-brown oblique stripes; head with bronze spots; fins yellow bronze with dusky margins; may display irregular dark bars and blotches on sides. SIZE: Maximum adult size about 46 cm (1.5 ft) TL; common to 30 cm (1 ft) TL. RANGE: Coastal waters from New York to Florida and throughout the Gulf of Mexico.

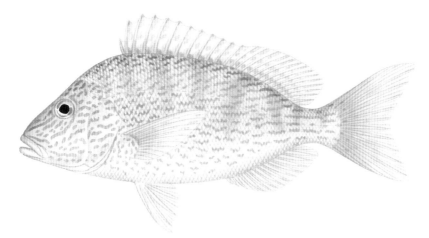

Pigfish *Orthopristis chrysoptera*

HABITAT AND HABITS: Pigfish primarily inhabit shallow tropical and subtropical waters and often frequent brackish water. They are bottom dwellers, most commonly over mud bottoms and occasionally over sandy vegetated areas. The common name for this family of fishes (grunts) derives from their habit of grinding the pharyngeal teeth, which causes resonation of the swim bladder. **OCCURRENCE IN THE CHESAPEAKE BAY:** Pigfish are occasional to common in the Chesapeake Bay from spring to autumn and more common in the lower than the upper bay, extending as far north as the Potomac River on the Western Shore and the Chester River on the Eastern Shore. Pigfish enter the bay after spawning occurs in spring and early summer. They leave the bay in late fall and apparently overwinter in deeper offshore waters.

REPRODUCTION: Spawning occurs offshore in the Mid-Atlantic Bight in spring and early summer.

FOOD HABITS: Pigfish form schools during the day and then disperse at night to feed on a variety of bottom-dwelling invertebrates, including worms, mollusks, amphipods, shrimps, and crabs. Small fishes are also sometimes eaten.

IMPORTANCE: Pigfish are of minor commercial and recreational importance in the Virginia portion of the bay. In the Chesapeake Bay area, pigfish are landed with seines, gillnets, and pound nets. In the 10-year period from 1999 to 2008, a total of 0.4 mt (940,000 lb) was landed commercially, all from Virginia waters. Anglers catch pigfish using bottom rigs baited with bloodworms, shrimps, or squids. This species is highly esteemed as a food fish and is also important as a bait fish. Virginia anglers caught more than 86,000 pigfish in 2004, the highest total in recent years.

Flier - *Centrarchus macropterus* (Lacepède, 1801)

KEY FEATURES: Body short, deep, very compressed; dorsal-fin base long, with 11–12 (rarely 13) spines and 12–15 soft rays; margin of dorsal fin broadly rounded posteriorly; caudal fin slightly emarginated; anal fin with 6–8 spines (typically 7 or 8) and 13–16 soft rays, its margin broadly rounded posteriorly; pelvic fin with long spine adjacent to long filamentous ray. **COLOR:** Olive green or brown over entire body, darker on back than belly; each scale with brown spot forming longitudinal rows of dots; wedge-shaped dusky bar courses through eye and onto cheek; dorsal and anal fins with numerous pale spots that form bands; smaller individuals with prominent orange and black ocellus (eyespot) near posterior lower margin of dorsal fin. **SIZE:** Maximum adult size approaches 20 cm (8 in) TL. **RANGE:** Mississippi Valley from Indiana and Illinois south to Louisiana and along the Gulf coast to the Florida panhandle. Along the Atlantic coast, fliers extend from north-central Florida to southern Maryland.

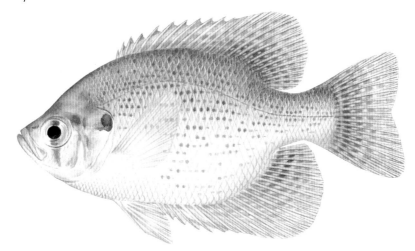

Flier *Centrarchus macropterus*

HABITAT AND HABITS: A typical flier habitat is an acidic well-vegetated slow-moving water body such as a pond, swamp, or sluggish stream. Fliers are typically bottom dwellers found near or around stumps, brush, or vegetation. **OCCURRENCE IN THE CHESAPEAKE BAY:** Fliers are locally common in some small coastal-plain streams in Virginia from the Rappahannock River drainage southward, but they are uncommon northward. In Maryland, they are known from tributaries of the lower Potomac River. They will occasionally penetrate into Chesapeake Bay waters of 5‰ or greater salinity and can tolerate salinities as great as 7‰.

REPRODUCTION: As with most centrarchids, the male constructs a nest in sand or gravel by fanning out a roundish depression in the substrate using the fins. One or more females will be enticed to lay eggs over the nest, and the male fertilizes them. The nest is guarded vigorously by the male until the young disperse. Spawning occurs in the spring when water temperatures reach about 17°C (63°F), typically between late March and early May. The maximum age of fliers is approximately seven years.

FOOD HABITS: Fliers feed on aquatic insects such as caddisflies, midges, and mayflies, crustaceans such as copepods, and other fishes, such as young bluegill.

IMPORTANCE: Of no commercial or recreational importance; however, they do make attractive aquarium fish.

Blackbanded sunfish - *Enneacanthus chaetodon*
(Baird, 1855)

Bluespotted sunfish - *Enneacanthus gloriosus*
(Holbrook, 1855)

Banded sunfish - *Enneacanthus obesus* (Girard, 1854)

KEY FEATURES: Differentiated by body shape and relative heights of spinous and soft parts of dorsal fin. Blackbanded sunfish deep and short-bodied, with spinous portion of dorsal fin slightly greater in height than soft-rayed portion. Bluespotted sunfish and banded sunfish have moderately deep and elongate bodies, but spinous portion of dorsal fin lower in height than soft-rayed portion in bluespotted sunfish, while spinous portion of dorsal fin about same height as soft-rayed portion in banded sunfish. COLOR: All three with black spot on gill cover; blackbanded sunfish have six to eight black bars, including one through eye; adult bluespotted sunfish have iridescent blue spots forming irregular patterns and no vertical bars on body; banded sunfish greenish or olive with several dusky bars on sides; young bluespotted sunfish also have bars; head profile between snout and first dorsal fin tends to be round and convex in banded sunfish but straight in bluespotted sunfish. SIZE: Maximum size is 10 cm (4 in) TL for blackbanded sunfish, 9 cm (3.5 in) TL for banded sunfish, and 8.5 cm (3.4 in) TL for bluespotted sunfish. RANGE: All found in fresh and brackish waters as far south as Florida. Blackbanded and bluespotted sunfishes also found on the Gulf coast of Florida, with bluespotted sunfish as far west as Biloxi Bay, Mississippi. Banded sunfish found as far north as New Hampshire, with bluespotted sunfish extending to New York State and blackbanded sunfish only as far north as New Jersey.

HABITAT AND HABITS: All three are bottom dwellers in heavily vegetated nutrient-poor acidic waters. Banded sunfish have less affinity for brackish water than blackbanded and bluespotted sunfish. OCCURRENCE IN THE CHESAPEAKE BAY: All found in waters of Maryland's Eastern Shore. Bluespotted sunfish are common in Western Shore waters of both Maryland and Virginia. Blackbanded sunfish are known from three counties in Virginia (Prince George, Surrey, and Sussex) but are not reported from any Chesapeake Bay tributaries. Banded sunfish are known from the James and York River drainages.

REPRODUCTION: For all three, spawning likely occurs in the bay region from early May to late June. Spawning and nest-building occur both on the bottom and in vegetation.

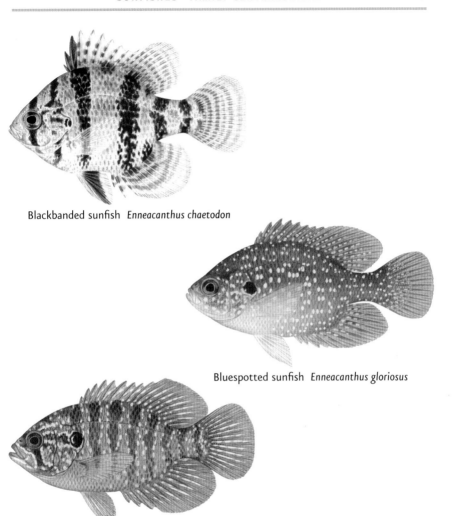

Blackbanded sunfish *Enneacanthus chaetodon*

Bluespotted sunfish *Enneacanthus gloriosus*

Banded sunfish *Enneacanthus obesus*

FOOD HABITS: Blackbanded sunfish are apparently nocturnal feeders on insects (primarily midge larvae) and some plant materials. Bluespotted sunfish feed primarily on crustaceans but also eat insects and worms. Small crustaceans and insects are the diet of banded sunfish.

IMPORTANCE: Due to their small size, these fishes are of no commercial or recreational importance, but they are handsome aquarium fishes.

Redbreast sunfish - *Lepomis auritus* (Linnaeus, 1758)

KEY FEATURES: Body deep and highly compressed, but rather elongate for a *Lepomis*; dorsal fin continuous, spinous portion arched, with 9–11 spines, soft portion rounded, with 10–12 soft rays; caudal fin slightly forked, with rounded lobes; pectoral fin short and rounded. **COLOR:** Blue green to olive back with sides bluish to greenish and belly yellow to whitish; scales with dusky centers and pale margins; gill flap black and very long in adults; snout and cheek with wavy blue lines; breast yellowish in females and reddish orange in males. **SIZE:** Maximum adult size is about 30 cm (1 ft) TL; however, the typical size for adults is 13–20 cm (5–8 in) TL. **RANGE:** Originally distributed from New Brunswick, Canada, to northern Florida. Redbreast sunfish have been introduced in many central and midwestern states.

Redbreast sunfish *Lepomis auritus*

HABITAT AND HABITS: Juvenile redbreast sunfish are often found in standing waters and the slower parts of streams, whereas adults are more commonly encountered in creeks and rivers. This sunfish will seek cover under rocks and may be associated with rock bass and smallmouth bass. Redbreast sunfish are typically a solitary species, but they will sometimes form small congregations at lower water temperatures. **OCCURRENCE IN THE CHESAPEAKE BAY:** These sunfish are common to abundant in all tributaries of the Chesapeake Bay. Redbreast sunfish are only occasionally encountered in brackish waters with salinities of 5‰ or greater; 7‰ is their maximum recorded salinity tolerance.

REPRODUCTION: Spawning in Maryland and Virginia waters occurs in June and July among plants near the shoreline; nests are scooped out of gravel bottoms. Redbreast sunfish have been observed carrying odd-sized stones away from the nest site in order to create a nest with similar-sized stones.

FOOD HABITS: Redbreast sunfish feed on a variety of invertebrate organisms, including insects and plankton, but they also eat fishes and crayfishes. Juveniles feed mainly on midges and small crustaceans such as ostracods. Adults feed on aquatic insects such as mayflies and dragonflies as well as fishes and crayfishes. They are also known to eat terrestrial insects such as beetles that may inadvertently enter the waterway.

IMPORTANCE: Because they are the most abundant sunfish in tidewater streams, redbreast sunfish are an important recreational species and can be taken with a fly rod as well as with spinning lures and baited hooks. They are excellent pan-fish. It is reported that anglers may catch these fish at night, which is unusual for sunfish.

Pumpkinseed - *Lepomis gibbosus* (Linnaeus, 1758)

KEY FEATURES: Body deep and highly compressed; dorsal fin continuous, spinous portion arched, with 9–12 (typically 10) spines, and soft portion rounded, with 10–13 soft rays; caudal fin slightly forked, with rounded lobes; pectoral fin long and sharply pointed. **COLOR:** Olive green to bluish with many irregular spots of bright copper or gold; breast orange, yellow, or gold; blue and orange streaks radiate from mouth onto cheeks; gill flap mostly black, but with spot of orange or red at rear edge. **SIZE:** Maximum adult size is about 31 cm (1 ft) TL; however, the typical size for adults is 10–15 cm (4–6 in) TL. **RANGE:** Known from Quebec to Georgia and as far west as North Dakota and Iowa. Pumpkinseed have been widely introduced elsewhere in the United States.

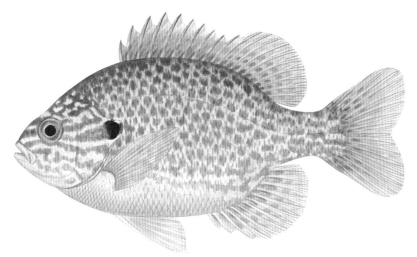

Pumpkinseed *Lepomis gibbosus*

HABITAT AND HABITS: Pumpkinseed typically occupy well-vegetated natural lakes, coastal ponds, and pools of rivers. This species can also tolerate acidic, darkly stained waters. Pumpkinseed are more common in still waters. **OCCURRENCE IN THE CHESAPEAKE BAY:** Pumpkinseed are common to abundant in all tributaries of the Chesapeake Bay and are frequently found in brackish waters with salinities greater than 5‰. They are tolerant of salinities as great as 18‰ for short periods.

REPRODUCTION: Spawning behavior is similar to bluegills and takes place once waters warm to about 20°C (68°F) in late April to early May, extending until August. Males construct nests over sandy or gravel bottoms in shallow waters, often near weed beds. Spawning may occur more than once a season. Pumpkinseed will hybridize with other sunfishes, such as bluegills and redbreast sunfish.

FOOD HABITS: Pumpkinseed are daytime feeders on a wide variety of small animals, including worms, mollusks, insects, and small fishes. They can live as long as 10 years.

IMPORTANCE: Well known for their beauty and tastiness and sought by anglers of all ages. Pumpkinseed will take natural and artificial baits and are easily caught near shore during the spawning season.

Bluegill - *Lepomis macrochirus* Rafinesque, 1819

KEY FEATURES: Body deep and highly compressed; dorsal fin continuous, spinous portion arched with 9–12 (typically 10) spines, with soft portion rounded, and 9–13 soft rays; caudal fin slightly forked, with rounded lobes; pectoral fin long and pointed. **COLOR:** Sides and back olive green, belly yellowish; purple-blue sheen sometimes evident on head; color extremely variable, may be completely dark during breeding, breeding males have reddish-orange breast, large males have hump on nape; six to eight olive chainlike vertical bars on sides; adults with black spot on lower half of last four rays of soft dorsal fin; gill flap bluish black. **SIZE:** Maximum adult size is about 30 cm (1 ft) TL; however, the typical size for adults is 10–15 cm (4–6 in) TL. **RANGE:** Introduced to the Chesapeake Bay region, probably in the mid–1880s. Originally distributed from the Great Lakes and Mississippi Valley to Texas and Florida. Bluegills have been widely introduced elsewhere in the United States as well as in Europe, Japan, and South Africa.

Bluegill *Lepomis macrochirus*

HABITAT AND HABITS: Bluegills are apt to be found in quiet waters of lakes, ponds, and slow-flowing rivers and streams bordered with aquatic vegetation, with sand, mud, or gravel bottoms. They are frequently stocked in ponds, along with largemouth bass. **OCCURRENCE IN THE CHESAPEAKE BAY:** This species is common to abundant in all tributaries of the Chesapeake Bay and can tolerate salinities at least as great as 18‰.

REPRODUCTION: Spawning in the Chesapeake Bay region is from April to September when water temperatures reach about 12°C (54°F). Multiple spawning is common. Males build nests in colonies where many nests are close together. The circular nest, 20–30 cm (8–12 in) in diameter, is constructed in shallow waters (usually less than 1 m, or 3.3 ft, deep) and consists of a shallow depression in a sandy or fine-gravel bottom from which the detrital covering has been fanned away. The male vigorously guards the nest both before and after spawning.

FOOD HABITS: Bluegills feed throughout the water column on a wide variety of organisms, including insects, crayfishes, snails, and other fishes. Bluegills are a common food item of larger fishes such as largemouth bass.

IMPORTANCE: Popular with anglers of all ages because of their readiness to take a baited hook. Artificial lures and flies are also used to catch bluegills. They are considered excellent food fish.

Smallmouth bass - *Micropterus dolomieu* Lacepède, 1802

KEY FEATURES: Mouth moderate, with upper jaw not extending beyond eye; 67–85 (typically 68–78) small scales in lateral series; caudal fin notched; anal fin shaped like soft dorsal fin, with 3 short graduated spines and 9–12 (typically 11) soft rays. **COLOR:** Uniformly olive green or pale brown, often with numerous small dusky brown blotches and sometimes with 5–15 indistinct dusky bars on sides; 3–5 brownish stripes on head below and behind eye; juveniles with tricolored caudal fin with orange base, black middle, and white margin. **SIZE:** Typical adult size is 20–56 cm (8–22 in) TL; maximum adult size approaches 70 cm (2.3 ft) TL. **RANGE:** The original range of smallmouth bass extended from the Great Lakes south to northern Georgia and Alabama, east to the Appalachian range, and west to eastern Oklahoma. Smallmouth bass have been widely introduced outside of this range in the northeastern United States, including the Chesapeake Bay region, and in areas west of the Mississippi drainage.

Smallmouth bass *Micropterus dolomieu*

HABITAT AND HABITS: The optimal smallmouth bass habitat is a cool clear stream with abundant shade and cover and with deep pools, moderate currents, and a gravel or rubble substrate. Smallmouth bass are also found in large clear lakes and reservoirs containing rocky shoals. Smallmouth bass exhibit strong cover-seeking behavior and prefer protection from light in all life stages. **OCCURRENCE IN THE CHESAPEAKE BAY:** Smallmouth bass tolerate salinities to about 7‰ and are occasional to common in Chesapeake Bay tributaries from the Pamunkey River (York River drainage) northward, rare to occasional south of there, and absent from Eastern Shore streams and rivers.

REPRODUCTION: Nest building and spawning occur from spring to early summer, usually mid-April to July, on rocky lake shoals or river shallows when water temperatures reach 13–21°C (55–70°F). The maximum known age for smallmouth bass is 18 years, but individuals older than 7 years are uncommon.

FOOD HABITS: Juvenile smallmouth bass eat large insects, crayfishes, and fishes, whereas adults feed primarily on fishes and crayfishes. Peak feeding activities occur at dawn and dusk.

IMPORTANCE: Known as important game fish that will take live bait, lures, or artificial flies. Because they only occasionally enter brackish waters, smallmouth bass are not considered a valuable species by Chesapeake Bay commercial fisheries.

Largemouth bass - *Micropterus salmoides*
(Lacepède, 1802)

KEY FEATURES: Mouth very large, with upper jaw extending beyond eye; 51–77 (typically 60–68) moderate scales in lateral series; caudal fin notched; anal fin shaped like soft dorsal fin, with 3 short graduated spines and 10–12 (typically 11) soft rays. **COLOR:** Dull green along back, yellowish on sides, white or cream-colored on belly; three dusky bars on head behind eye; midlateral stripe, dark green to blackish green, present from snout to caudal fin, which becomes less distinct or blotchy as fish age, juveniles lack midlateral stripe. **SIZE:** Maximum adult size is about 88 cm (2.9 ft) TL. The maximum known age for largemouth bass is 25 years. **RANGE:** Native to the central and eastern United States, excluding the northeastern and central Atlantic states. Largemouth bass have been widely introduced elsewhere in North America and the world. This species was introduced to the Chesapeake Bay region in the mid–1800s.

Largemouth bass *Micropterus salmoides*

HABITAT AND HABITS: Optimal habitat for largemouth bass are lakes with extensive shallow areas (<6 m, or 20 ft, deep) that support submerged vegetation. Largemouth bass are also found in large slow-moving rivers or pools of streams with soft bottoms, some aquatic vegetation, and relatively clear water. OCCURRENCE IN THE CHESAPEAKE BAY: Largemouth bass are common to abundant in all tributaries of the Chesapeake Bay and can tolerate salinities as high as 24‰.

REPRODUCTION: Spawning occurs in the spring when water temperatures are 12–18°C (54–64°F). A gravel substrate is preferred for spawning, but largemouth bass will nest on a wide variety of other substrates, including vegetation, roots, sand, mud, and rocks. Nests are constructed by the male at depths less than 1 m (3.3 ft).

FOOD HABITS: Largemouth bass juveniles consume mostly insects and small fishes, whereas adults feed primarily on fishes and crayfishes. Largemouth bass are also known to eat snakes, frogs, and ducklings. Adults often feed near vegetation within shallow areas; peak feeding is in early morning and late evening.

IMPORTANCE: One of the most sought-after game fishes in the United States, particularly throughout the Atlantic seaboard, Mississippi Valley, and tributaries of the Gulf of Mexico. In tributaries of the Chesapeake Bay, they are important food and game fish but not abundant enough in brackish waters to be of value to the commercial fisheries of the bay.

Black crappie - *Pomoxis nigromaculatus* (Lesueur, 1829)

KEY FEATURES: Body deep and strongly compressed, with large oblique mouth; dorsal fin continuous, with 6–10 spines (typically 7 or 8); soft portion of dorsal fin high and rounded, with 14–16 rays; caudal fin slightly forked, with blunt lobes. COLOR: Dark green on back, with patches of dusky scales coalescing to form irregular blotches; small scattered dusky blotches along sides, forming no discernible pattern; dorsal, anal, and caudal fins with greenish spots. SIZE: Maximum adult size is about 49 cm (1.6 ft) TL. RANGE: Known in freshwater lakes and streams from the St. Lawrence Valley west to Manitoba, south to Texas and Florida. Black crappies were introduced to the Chesapeake Bay region in the 1860s and have been widely introduced elsewhere in North America.

HABITAT AND HABITS: Black crappies are most abundant in well-vegetated lakes and clear backwaters of rivers. This species is often found associated with submerged tree limbs, brush piles, and fallen trees. OCCURRENCE IN THE CHESAPEAKE BAY: Black

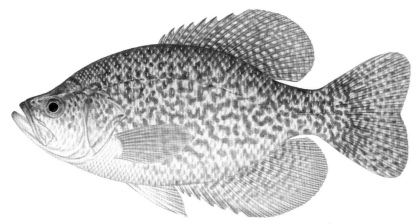

Black crappie *Pomoxis nigromaculatus*

crappies are occasional to abundant in all major tributaries of the Chesapeake Bay and can tolerate salinities at least as great as 5‰.

REPRODUCTION: Spawning occurs from March to July when water temperatures are 15–20°C (59–68°F). Male black crappies construct and guard nests in sandy bottoms of weedy areas at depths of 1.0–2.5 m (3.3–8.2 ft). The maximum age for black crappies is about 10 years.

FOOD HABITS: Young black crappies are plankton feeders, but adults feed primarily on small fishes and will also eat benthic insects and crustaceans. This species feeds at night or in early morning.

IMPORTANCE: Black crappies are excellent game and food fish that are pursued by many freshwater anglers. This species will take a hook in late winter or early spring before other fishes begin to feed. They are only rarely encountered in brackish waters and are of no commercial importance.

Tautog - *Tautoga onitis* (Linnaeus, 1758)

KEY FEATURES: Lips broad and thick; teeth in jaws strong, anterior ones compressed and incisor-like; dorsal fin long and continuous, with 16–18 spines and 10–11 soft rays, soft rays longer than spines. **COLOR:** Dull black to greenish black or brown dorsally, with irregular blackish lateral bars or blotches. Small males often uniformly blackish. Most large males have enlarged white chin and white dorsal and ventral margins on pectoral and caudal fins. Juveniles less than about 5 cm (2 in) TL often bright green. **SIZE:** Maximum adult size is 95 cm (3.2 ft) SL. **RANGE:** Known from the Bay of Fundy to South Carolina; most abundant from Cape Cod to Virginia.

adult male

juvenile/female

Tautog *Tautoga onitis*

HABITAT AND HABITS: Tautog are coastal fishes that commonly associate with rock piles, bridge pilings, artificial reefs, and old wrecks, much like sheepshead do. **OCCURRENCE IN THE CHESAPEAKE BAY:** Tautog are year-round residents near the Chesapeake Bay mouth and extend seasonally as far north as the Chester River. Tautog are locally abundant in the lower bay from autumn to spring. A population shift to more offshore locations occurs in summer and perhaps again in January and February.

REPRODUCTION: Spawning occurs from late April to early August both in the lower bay and offshore. Young are planktonic for about three weeks and then take up residence in shallow seagrass beds, with which the green color of the young blends well. Tautog mature at three years of age in the Chesapeake Bay region. They are slow-growing fish known to live as long as 34 years. Hermaphroditism has not been established for tautog; however, males are dimorphic, with some having an enlarged white chin, as described above, and others, usually smaller males, being blackish and similar to females. The sex ratio is known to be strongly skewed toward males in older fishes.

FOOD HABITS: Tautog feed on a variety of mollusks and crustaceans such as mussels, barnacles, and crabs, which are crushed by its strong molars.

IMPORTANCE: Tautog are of minor commercial value in the Chesapeake Bay region, with total landings rarely exceeding 8 mt (17,600 lb) in recent years. However, they are very popular with anglers, who take them by bottom-fishing with hook and line baited with crabs or clams. Tautog are also a popular target of spearfishers. In the lower bay, tautog are most available around wrecks and other submerged structures in spring and fall. Recreational catches of tautog from waters of the Chesapeake Bay region have averaged nearly 90,000 fish annually in recent years.

Cunner - *Tautogolabrus adspersus* (Walbaum, 1792)

KEY FEATURES: Lips thin; outer row of teeth enlarged and canine-like, protruding anteriorly; dorsal fin long and continuous, with 16–18 spines and 9–11 soft rays, soft rays longer than spines. **COLOR:** Highly variable, may be uniformly brownish to olive green or blue, sometimes mottled; body color frequently matching that of bottom they inhabit. Fins reddish, dusky spot on base of anterior soft rays. **SIZE:** Maximum adult size about 41 cm (1.3 ft) TL. **RANGE:** Known from Labrador to Virginia but more common north of New Jersey and very abundant in the Massachusetts Bay region.

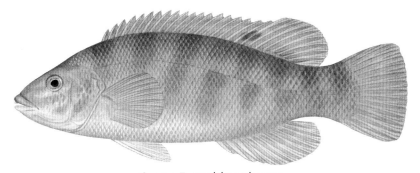

Cunner *Tautogolabrus adspersus*

HABITAT AND HABITS: Cunner live between the shoreline and depths as great as 123 m (403 ft). However, they are most common inshore, particularly in bays and sounds, and seldom move far from a rocky outcrop or similar structures. In the southern part of their range, cunner are found in slightly deeper waters. Larger cunner tend to be found in deeper water than smaller ones. As with other wrasses, cunner are active in daytime and spend the night hours in rock crevices or resting alongside a structure, and swim using only their pectoral fins. This creates the impression that they are dragging their tails; only when a burst of speed is needed do cunner use their tails for locomotion. **OCCURRENCE IN THE CHESAPEAKE BAY:** Cunner are rare to occasional visitors to the lower Chesapeake Bay during summer, autumn, and winter.

REPRODUCTION: Spawning in the mid-Atlantic region takes place from near shore to the edge of the continental shelf from May to November, with a peak in June and July.

FOOD HABITS: Cunner feed both on the bottom and in the water column on a wide variety of animals, including crabs, shrimps, bivalves, and worms. Juvenile cunner feed in the water column on swimming crustaceans, while older fish tend to target mussels and barnacles.

IMPORTANCE: Cunner are of no commercial or recreational importance in the Chesapeake Bay region. However, catches of cunner are occasionally reported by anglers in Virginia's bay waters. Cunner have excellent eating qualities, and where cunner are abundant, they are caught frequently by anglers. Cunner are well known for their bait-stealing ability.

Barrelfish - *Hyperoglyphe perciformis* (Mitchill, 1818)

KEY FEATURES: Dorsal fin continuous, with 6–8 very short stout spines followed by 19–23 soft rays, spinous portion of dorsal fin much lower than rayed portion. **COLOR:** Adults brownish black dorsally, grading to brownish white ventrally. Some individuals collected near the surface have greenish flecks laterally and ventrally. **SIZE:** Maximum adult size about 1 m (3.3 ft) TL. **RANGE:** In the western Atlantic, barrelfish are known from Nova Scotia to Key West, Florida. There are reports of barrelfish in the eastern Atlantic from Ireland, Spain, and Portugal.

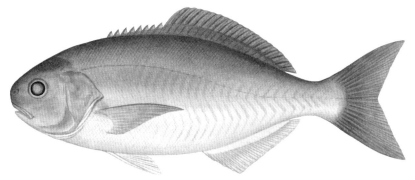

Barrelfish *Hyperoglyphe perciformis*

HABITAT AND HABITS: Barrelfish owe their common name to their habit of congregating around floating wood, planks, or wreckage and on occasion, drifting inside boxes in flotsam. Near-surface drifting is primarily a habit of the young fish (5–10 cm, or 2–4 in, TL), which are also found under floating seaweed. Adult barrelfish are a deep-dwelling species over the continental slope and in submarine canyons. **OCCURRENCE IN THE CHESAPEAKE BAY:** This species is a rare visitor to the Chesapeake Bay during summer and autumn. Only juvenile barrelfish are likely to be observed in the bay, because juveniles typically drift close inshore.

REPRODUCTION: No information is available on spawning.

FOOD HABITS: Young barrelfish feed on hydroids, tunicates, salps, ctenophores, and barnacles as well as young fishes of various kinds. Adult fish are known to feed on squid.

IMPORTANCE: Of no commercial or recreational importance anywhere. Barrelfish are caught incidentally in trawls, in nets, and on baited hooks off the coast of the United States. The world record barrelfish was caught off the Virginia Beach coast in November 2008 and weighed 9.5 kg (20.9 lb).

Black sea bass - *Centropristis striata* (Linnaeus, 1758)

KEY FEATURES: Dorsal fin continuous but notched, with 10 slender spines and 10–12 soft rays; in adults, spine tips with fleshy filaments; anal fin with 3 spines and 7 soft rays; caudal fin rounded to trilobate, large individuals usually with one of the upper rays elongate. **COLOR:** Body bluish black in adults, brownish in juveniles; centers of scales pale blue or white, forming longitudinal lines along back and sides; dorsal fin with longitudinal rows of pale streaks, females paler than males; juveniles with prominent dusky broad lateral stripe and dusky blotch at base of posterior dorsal spines; adult males with bright blue marking on head. **SIZE:** Maximum adult size is about 60 cm (2 ft) TL. **RANGE:** In the western Atlantic, black sea bass are known from Massachusetts to central Florida. Black sea bass are also found on the gulf coast of Florida from Pensacola to Placida.

Black sea bass *Centropristis striata*

HABITAT AND HABITS: Adult black sea bass are most often found on rocky bottoms near pilings, wrecks, and jetties. Juveniles are found in vegetated flats and channels.

OCCURRENCE IN THE CHESAPEAKE BAY: Black sea bass are common in the Chesapeake Bay from spring to late autumn, extending as far north as the Chester River. In the winter, they migrate offshore and south. Large fish are more common offshore than in the bay.

REPRODUCTION: Black sea bass are protogynous hermaphrodites—that is, initially they are females, but larger fish reverse sex to become males. Spawning begins in June, peaks in August, and continues through October in offshore waters in the mid-Atlantic region. During the spawning season, males develop a fleshy hump on the nape that is often bright blue. Black sea bass are reported to live as long as 20 years. However, individuals longer than 38 cm (1.3 ft) TL, the size of an approximately 8-year-old fish, are uncommon.

FOOD HABITS: Black sea bass are visual feeders during daylight hours. The adults feed chiefly on crabs, mussels, razor clams, and fishes, whereas the young prey on shrimp, isopods, and amphipods.

IMPORTANCE: Commercial landings of black sea bass from the Chesapeake Bay region are significant, averaging about 340 mt (749,700 lb) per year during the period 1999–2008, with the majority of landings coming from Virginia waters. This species also forms the basis of an important sport fishery. An average of 41,000 black sea bass are taken each year by fishers in bay waters of Maryland and Virginia. Bottom-fishing from boats, anglers use squids and other natural baits fished over firm bottoms near rocks, wrecks, and reefs to catch these highly esteemed and flavorful fish.

Gag - *Mycteroperca microlepis* (Goode and Bean, 1879)

KEY FEATURES: Small sharp canine teeth at front of each jaw; dorsal fin continuous, with 11 (rarely, 10 or 12) slender spines and 16–19 soft rays; caudal fin truncate. **COLOR:** Body color variable and dependent on age and sex. Adult females and juveniles are pale to brownish gray, with prominent worm-like markings on sides. Mature males are paler gray with fainter worm-like markings. Some individuals almost entirely black. **SIZE:** Maximum adult size is about 120 cm (3.9 ft) TL. **RANGE:** Adult gag are found from North Carolina to the Yucatán Peninsula. This species is also reported from Brazil. Juveniles inhabit estuaries as far north as Massachusetts.

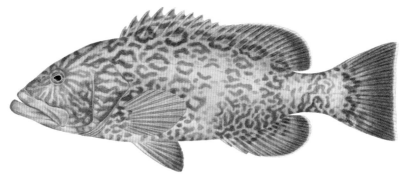

Gag *Mycteroperca microlepis*

HABITAT AND HABITS: Adults frequent subtropical and tropical waters from 20 to 80 m (66 to 262 ft) in depth in rocky areas and on coral reefs. Gag may be found alone or in small schools of a few dozen individuals. Juveniles occur in estuaries and seagrass beds. **OCCURRENCE IN THE CHESAPEAKE BAY:** Juvenile gag are occasional visitors to seagrass beds in the lower Cheasapeake Bay during summer and early autumn. Individuals in the size range 7–22 cm (3–9 in) have been collected from both the Eastern and Western Shores of the bay in Virginia waters.

REPRODUCTION: Gag are protogynous hermaphrodites, with sexual transition from female to male occurring between 10 and 11 years of age. Spawning occurs off the southeast Atlantic coast in February at depths greater than 70 m (230 ft). Gag may live for 15 years.

FOOD HABITS: Adult gag feed mainly on fishes, but crabs, shrimps, and squids are also included in the diet. Juveniles prey primarily on crustaceans.

IMPORTANCE: Gag are of no commercial or recreational importance in the Chesapeake Bay, because of scarcity. Gag are occasionally taken in pound nets or by anglers bottom-fishing near the bay mouth.

Gray triggerfish - *Balistes capriscus* Gmelin, 1789

KEY FEATURES: Head and body deep, strongly compressed; scales of moderate size, implanted in thick leathery skin; dorsal fins separate, the first comprising three short strong spines, the first of which is longest; dorsal and ventral caudal-fin rays prolonged; pelvic fins represented by single multi-barbed blunt spiny process at terminus of pelvic bone. COLOR: Grayish green with three dusky blotches above; small purplish or bluish spots on dorsal part of body, pale spots on belly; second dorsal and anal fins with dusky streaks. SIZE: Maximum adult size about 40 cm (1.3 ft) TL. RANGE: Known from both the eastern and western Atlantic coasts. In the western Atlantic, gray triggerfish range from Nova Scotia to Argentina, including the Gulf of Mexico and Caribbean Sea.

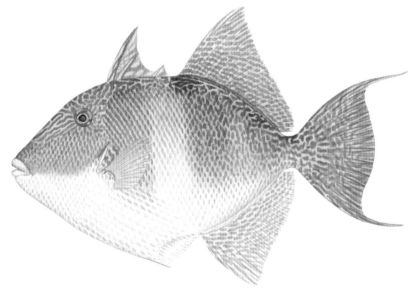

Gray triggerfish *Balistes capriscus*

HABITAT AND HABITS: Gray triggerfish frequent coral reef environments, including sand or grass flats, as well as rocky bottoms, to depths of about 50 m (165 ft). Adults are demersal and swim along the bottom in small groups or alone, whereas juveniles often associate with floating seaweed. The first dorsal-fin spine of triggerfishes can be locked in place by the second spine (the "trigger"). At night or when alarmed, these fish take refuge in a rock or coral crevice, lock their dorsal spine upward, and erect their pelvic bone, thus becoming effectively wedged in and protected from attack. They swim by undulating the second dorsal and anal fins, using the tail only to move quickly. **OCCURRENCE IN THE CHESAPEAKE BAY:** Gray triggerfish are occasional summer visitors to the lower bay, reaching as far north as the mouth of the Potomac River.

REPRODUCTION: Spawning occurs from July to September and only after water temperatures exceed 21°C (70°F). Eggs are laid in a shallow depression made by the female in sand bottoms, and the nest is aggressively guarded by the male.

FOOD HABITS: Gray triggerfish have powerful jaws and chisel-like teeth, with which they feed on a variety of invertebrate organisms, including crabs, mollusks, sea urchins, and coral.

IMPORTANCE: Gray triggerfish are not of commercial or recreational importance in the Chesapeake Bay due to lack of abundance. The total annual catch of gray triggerfish by anglers in Virginia's bay waters has averaged only about 4,500 fish in recent years.

Striped burrfish - *Chilomycterus schoepfii*
(Walbaum, 1792)

KEY FEATURES: Body covered with prominent spines; spines relatively short and massive, covering head and body. **COLOR:** Yellowish green dorsally, whitish orange ventrally; numerous dusky brown or blackish irregular wavy stripes on dorsum; several large black ocelli (eye-like spots) laterally on body, number is variable but commonly one above and another immediately behind pectoral fin, one below dorsal fin, and another just in front of caudal peduncle; fins greenish yellow or orange. **SIZE:** Maximum adult size about 25 cm (10 in) SL. **RANGE:** Nova Scotia to Belize, including the Gulf of Mexico, Cuba, and the Bahamas.

HABITAT AND HABITS: Striped burrfish are typically found in deep flats, in grass flats, and along channel margins. When disturbed, these fish inflate to an almost spherical shape, with their long spines directed outward. They swim slowly, propelled by un-

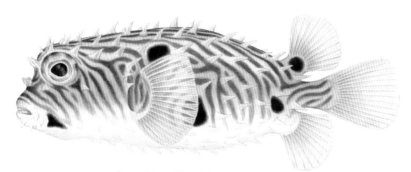

Striped burrfish *Chilomycterus schoepfii*

dulations of the dorsal and anal fins, and use their tail as a rudder. Their locomotion is reported to be assisted by jetting water from the gill openings. **OCCURRENCE IN THE CHESAPEAKE BAY:** Striped burrfish are common visitors to the lower to middle bay from late spring to autumn, reaching as far north as the Patuxent River. Striped burrfish move out of the bay and southward in winter.

REPRODUCTION: No information on the reproductive habits of striped burrfish is available.

FOOD HABITS: Striped burrfish feed primarily on benthic invertebrates, especially hermit crabs, which are consumed shell and all.

IMPORTANCE: Striped burrfish are of no commercial or recreational value. They are often dried in their inflated state and used for curio lampshades and souvenirs.

Orange filefish - *Aluterus schoepfii* (Walbaum, 1792)
Scrawled filefish - *Aluterus scriptus* (Osbeck, 1765)

KEY FEATURES: These two filefishes can be easily distinguished based on color, fin-ray counts, and body depth. Scrawled filefish have higher numbers of dorsal-fin rays (43–50) and anal-fin rays (46–52) than do orange filefish (32–40 and 35–41, respectively). Orange filefish are deeper bodied than scrawled filefish. COLOR: In scrawled filefish, the head and body are shades of olive, with blue to greenish reticulations and round blackish spots. Orange filefish are typically dirty orange to brownish, with a dense covering of orange to brownish spots on the head and body, and may display several very dark bars and blotches. Both species can change color rapidly. SIZE: Scrawled filefish reach a maximum adult size of about 90 cm (3 ft) TL, larger than the maximum adult size of orange filefish (60 cm, or 2 ft, TL). RANGE: Scrawled filefish are known from all tropical seas. In the western Atlantic, scrawled filefish range from Georges Bank to Brazil, including the Gulf of Mexico, rarely occurring north of South Carolina. Orange filefish range from Nova Scotia to Brazil, including the Gulf of Mexico and Caribbean Sea.

HABITAT AND HABITS: Orange filefish are a shallow-water species that typically associates with seagrass beds. Scrawled filefish typically inhabit patch reefs and grass beds at depths of 3–120 m (10–394 ft). Young scrawled filefish mimic blades of seagrass and are observed drifting head-down, looking at the bottom for food. OCCURRENCE IN THE CHESAPEAKE BAY: Scrawled filefish are rare summer visitors to the lower bay. Orange filefish are summer visitors to the Chesapeake Bay, common in the lower bay but only occasionally encountered in the middle bay, extending as far as the Choptank and Patuxent Rivers.

REPRODUCTION: There is no information available on either filefish other than that larvae and juveniles associate with floating seaweed.

FOOD HABITS: Bottom-dwelling organisms such as algae, seagrass, anemones, and tunicates form the bulk of the diet of both these filefishes.

IMPORTANCE: Neither species is of any commercial or recreational value. These filefishes are sometimes taken as bycatch in trawls.

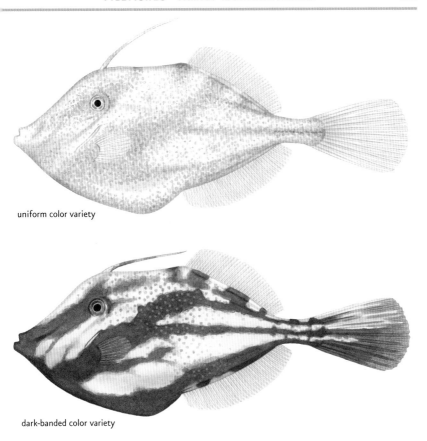

uniform color variety

dark-banded color variety

Orange filefish *Aluterus schoepfii*

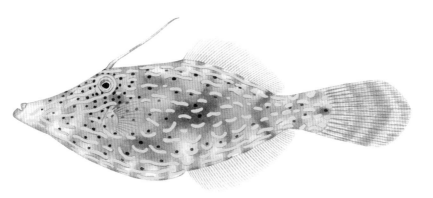

Scrawled filefish *Aluterus scriptus*

Planehead filefish - *Stephanolepis hispidus*
(Linnaeus, 1766)

KEY FEATURES: Scales minute with short rough bristles; first dorsal fin consisting of a single barbed spine originating dorsal to posterior part of eye; second dorsal-fin ray elongate in males; pelvic spine prominent, represented as a barbed process in midline at end of pelvis. **COLOR:** Color variable with background, typically grayish or greenish; sides with irregular grayish-brown blotches; caudal fin dusky. **SIZE:** Maximum adult size about 20 cm (8 in) TL. **RANGE:** Found in both the eastern and western Atlantic. Along the western Atlantic coast, planehead filefish are known from Nova Scotia to Uruguay and throughout the Gulf of Mexico.

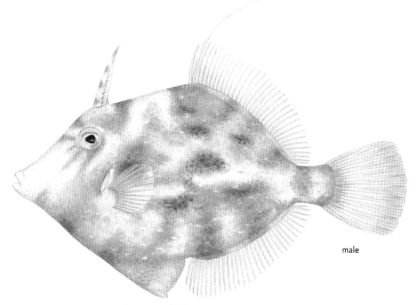

male

Planehead filefish   *Stephanolepis hispidus*

HABITAT AND HABITS: Adult planehead filefish are often associated with seagrass beds; juveniles occasionally associate with floating seaweed and pilings. **OCCURRENCE IN THE CHESAPEAKE BAY:** Planehead filefish are occasional visitors to the lower Chesapeake Bay during the summer and autumn.

REPRODUCTION: The male prepares a nest where the female will deposit her eggs, which are guarded until they hatch. Newly hatched larvae are planktonic.

FOOD HABITS: Planehead filefish feed on bryozoans, small crustaceans, mollusks, worms, sea urchins, and algae.

IMPORTANCE: Planehead filefish are of no commercial or recreational value.

Trunkfish - *Lactophrys trigonus* (Linnaeus, 1758)

KEY FEATURES: Body box-shaped and covered by a carapace with only jaws, fins, and tail free; carapace incomplete posteriorly, partially open posterior to dorsal fin. **COLOR:** Greenish to tannish, with small white spots and two dusky blotches on sides. May appear very dark. **SIZE:** Maximum adult size 45 cm (1.5 ft) TL. **RANGE:** From Massachusetts to Brazil, including the Gulf of Mexico and Caribbean Sea.

Trunkfish *Lactophrys trigonus*

HABITAT AND HABITS: As would be expected from their appearance, trunkfish are slow swimmers; they propel themselves primarily by a sculling action of the dorsal and anal fins. Trunkfish typically inhabit seagrass beds in shallow water to depths of 50 m (165 ft). **OCCURRENCE IN THE CHESAPEAKE BAY:** Rare summer visitors to the lower Chesapeake Bay. Typically only juvenile trunkfish enter the bay, as tropical strays. The only record from Maryland waters is a juvenile collected from St. Peter's Creek of the Manokin River, which is just south of the Nanticoke River mouth.

REPRODUCTION: No information is available.

FOOD HABITS: Trunkfish are omnivorous, feeding on a variety of small bottom-dwelling organisms such as sea grasses, tunicates, mollusks, crustaceans, and worms.

IMPORTANCE: Of no commercial or recreational importance. Trunkfish are caught incidentally with traps or by seining.

Smooth puffer - *Lagocephalus laevigatus* (Linnaeus, 1766)

KEY FEATURES: Ventral edge of body with longitudinal fold or keel; abdomen with short spines, elsewhere skin smooth; caudal-fin margin deeply concave, upper lobe slightly longer. COLOR: Back and upper sides uniformly dusky gray or greenish gray, lower sides silver, belly white; young fish with several broad dark bars on upper sides. SIZE: These are the largest of the American puffers, with a maximum adult size of 1 m (3.3 ft) TL. RANGE: Known from both the eastern and western Atlantic coasts. In the western Atlantic, smooth puffers range from Massachusetts to Argentina, including the Gulf of Mexico and Caribbean Sea.

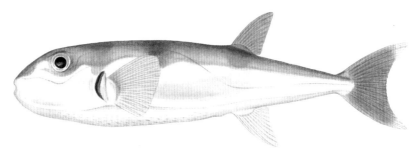

Smooth puffer *Lagocephalus laevigatus*

HABITAT AND HABITS: Smooth puffers inhabit inshore and nearshore areas to about 60 m (200 ft) in depth, over sand, rocky, or mud bottoms, and they are known to form small aggregations. OCCURRENCE IN THE CHESAPEAKE BAY: Smooth puffers are occasional summer and autumn visitors to the lower bay and are rarely encountered in the upper bay, extending as far as the mouth of the Chester River.

REPRODUCTION: Smooth puffers lay adhesive demersal eggs that are guarded by the male.

FOOD HABITS: Smooth puffers feed on benthic crustaceans and occasionally finfish.

IMPORTANCE: Smooth puffers are of no commercial or recreational importance. They are caught incidentally on hook and line while bottom-fishing. Because the smooth puffer has little or no toxicity, the flesh is edible and considered to be of excellent quality.

Northern puffer (pufftoad, blowfish) - *Sphoeroides maculatus* (Bloch and Schneider, 1801)

KEY FEATURES: Body with small spines or prickles except on caudal peduncle; caudal fin rounded. **COLOR:** Back grayish green with ill-defined black blotches and small black spots; sides yellowish, with series of several irregular blackish bars or blotches; intense black spot in pectoral-fin axil; belly white to yellow. **SIZE:** Maximum adult size 36 cm (1.2 ft) TL; common to 20–25 cm (8–10 in) TL. **RANGE:** Northern puffers occur coastally from Newfoundland to southern Florida.

Northern puffer *Sphoeroides maculatus*

HABITAT AND HABITS: Puffers are so named for their ability, when provoked, to greatly inflate themselves by drawing water (or air if out of the water) into a specialized chamber near the stomach. The resulting prickly ball deters many predators. Northern puffers are bottom dwellers that inhabit nearshore waters, including the surf zone as well as bays and estuaries. They are also found offshore to depths of 60 m (200 ft) or more. **OCCURRENCE IN THE CHESAPEAKE BAY:** Northern puffers are common visitors to the bay from spring to autumn, exiting the bay during winter. They are more common in the lower bay than in the upper, extending as far as Love Point, at the mouth of the Chester River.

REPRODUCTION: Spawning occurs from May to August in nearshore waters, where the eggs are attached to the substrate and guarded by the male.

FOOD HABITS: Northern puffers use their strong teeth to feed primarily on shellfish but will occasionally eat other fishes.

IMPORTANCE: Puffers are well known for possessing one of nature's most powerful toxins, tetrodotoxin, particularly in the liver and ovaries. Serious illness and death may result from eating these tissues. Muscle tissue of puffers is usually safe to eat and considered a delicacy by some people. Northern puffers, usually marketed as "sea squab," were an important food fish along the mid-Atlantic coast through the 1960s. During the period 1960–69, the annual commercial catch from Maryland and Virginia waters averaged 2,170 mt (4.8 million lb). Since that decade, the numbers have decreased dramatically and the northern puffer is now only an incidental catch for commercial fisheries in Maryland and Virginia, averaging only 21 mt (46,387 lbs) per year in recent years. In the bay, northern puffers are caught during the spring in pound nets and in the summer and autumn in crab pots. They are of minor recreational value to hook-and-line anglers fishing from boats or the shore; the average annual catch from waters of the Chesapeake Bay is about 10,000 individual fish.

Hogchoker - *Trinectes maculatus*
(Bloch and Schneider, 1801)

KEY FEATURES: Eyes small and on right side; dorsal- and anal-fin rays not connected to caudal fin; distinct caudal peduncle; pectoral fins usually absent (sometimes with single ray). COLOR: Color pattern highly variable, with ability to change rapidly (seconds to minutes). Most common pattern of ocular side (right side) brownish gray with seven or eight thin black vertical lines, but often with variable number of small to large spots. Blind side sometimes plain white or pale brown with variable number of dark spots, sometimes with same pattern as ocular side and occasionally partially pigmented and partially white. Fins mottled or with spots. SIZE: Maximum adult size about 20 cm (8 in) TL. RANGE: Known from Massachusetts to Florida and throughout the Gulf of Mexico to Venezuela.

HABITAT AND HABITS: Able to tolerate a wide range of salinities, hogchokers are found in tidal rivers and creeks as well as to depths as great as 25–75 m (82–246 ft) on the continental shelf. They are typically found on mud bottoms but can also inhabit sandy or silty bottoms. Larger individuals tend to be found closer to the mouth of the estuary and on the continental shelf. In addition to a wide range of salinity, hogchokers can tolerate low-oxygen conditions for short periods. OCCURRENCE IN THE CHESAPEAKE BAY: Hogchokers are abundant year-round residents found throughout the bay. They tend to ascend rivers and enter freshwater in winter.

REPRODUCTION: In the Chesapeake Bay region, spawning occurs from May to Sep-

Hogchoker *Trinectes maculatus*

tember in inshore waters and estuaries, with the majority of the spawning activity taking place at dusk. Young-of-the-year migrate upstream and congregate on shallow mud flats, where they overwinter. In spring, they move downstream.

FOOD HABITS: Hogchokers are lie-in-wait predators because of their ability to change colors to match their background. This species feeds on a variety of worms, crustaceans, and small fishes.

IMPORTANCE: Hogchokers are of no commercial or recreational importance and are considered trash fish. Hogs that are fed trash fish reportedly have great difficulty swallowing this sole, hence the common name "hogchoker."

Blackcheek tonguefish - *Symphurus plagiusa*
(Linnaeus, 1766)

KEY FEATURES: Eyes small, on left side of head; dorsal and anal-fin rays confluent with caudal fin; posterior two-thirds of body tapers to point; pectoral fins absent. COLOR: The common name for this fish derives from the large dusky spot that is usually present on the upper portion of the gill flap. The ocular side is brownish, sometimes with 6 or 7 indistinct broad crossbars formed by small dusky spots. Juveniles can have up to 10 such crossbars. The blind side is white. SIZE: Maximum adult size about 21 cm (8 in) SL. RANGE: From Connecticut to Florida and throughout the Gulf of Mexico as well as the Bahamas and Cuba. Adult blackcheek tonguefish are rare north of the Chesapeake Bay, home to the northernmost major population of this species.

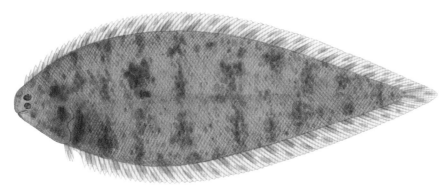

Blackcheek tonguefish *Symphurus plagiusa*

HABITAT AND HABITS: Blackcheek tonguefish have a wide salinity tolerance and can be found from freshwater to full seawater habitats. This species is most abundant over soft sediments such as mud but can be found in surf zones and on live bottoms. Offshore, blackcheek tonguefish appear to be more common over sand-silt sediments and reach depths of 40 m (130 ft). OCCURRENCE IN THE CHESAPEAKE BAY: Blackcheek tonguefish are year-round residents of the Chesapeake Bay, common in the lower bay but only occasionally encountered in the upper bay, extending as far as Kent Island just south of the Chester River. Juveniles are abundant in tidal creeks and saltmarsh fringes and move into deeper bay waters as they grow.

REPRODUCTION: Spawning occurs within the Chesapeake Bay from late spring through summer. South of the Chesapeake Bay, spawning habitats include shallow coastal and shelf waters as well as large estuaries. Spawning in these habitats occurs from March through October, with a peak in June and July.

FOOD HABITS: Blackcheek tonguefish feed on a diverse group of invertebrates such as mollusks, worms, and small crustaceans, as well as algae.

IMPORTANCE: These small tonguefish are of no commercial or recreational value and are taken incidentally with seines and bottom trawls.

Bay whiff - *Citharichthys spilopterus* Günther, 1862
Fringed flounder - *Etropus crossotus*
Jordan and Gilbert, 1882

KEY FEATURES: Both species have a nearly straight lateral line, 75–84 dorsal-fin rays, and 58–63 anal-fin rays. Bay whiffs have a much larger mouth than fringed flounder, while fringed flounder have a deeper body than bay whiffs. COLOR: The ocular side of both species is greenish brown, with or without dusky spots. In fringed flounder, the dorsal and anal fins have dusky spots and blotches, and the gill cover on the blind side has a distinct white fringe. SIZE: Both bay whiffs and fringed flounder attain a maximum adult size of about 20 cm (8 in) TL and are common to about 15 cm (6 in) TL. RANGE: Bay whiffs are known from New Jersey to Brazil, including the Gulf of Mexico and Caribbean Sea. They are rare north of Virginia. Fringed flounder range from the Chesapeake Bay to Brazil, including the Gulf of Mexico and parts of the Caribbean Sea.

HABITAT AND HABITS: Although reported from depths as great as 73 m (240 ft), bay whiffs are typically found in shallow waters with mud bottoms. Bay whiffs move offshore in winter. Fringed flounder inhabit mud, sand, and crushed-shell bottoms in depths to 65 m (215 ft) and are known to enter freshwater occasionally. OCCURRENCE IN THE CHESAPEAKE BAY: Bay whiffs are rare late-summer visitors to the lower Chesapeake Bay. Fringed flounder are rare to occasional visitors to the lower bay from spring to autumn and probably migrate southward during the winter months.

REPRODUCTION: Bay whiffs spawn in bays during late spring and early summer. Fringed flounder apparently spawn in late spring in the mid-Atlantic region.

FOOD HABITS: Juveniles of both species prey primarily on copepods, whereas adults feed on a variety of bottom-dwelling invertebrates, including shrimps.

IMPORTANCE: Due to their small size, neither species is of any commercial or recreational importance.

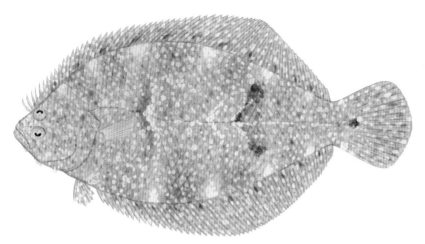

Bay whiff *Citharichthys spilopterus*

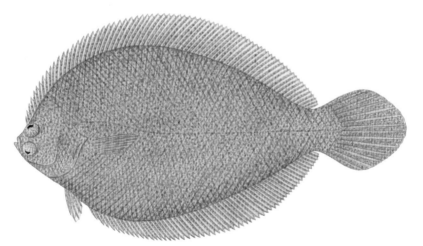

Fringed flounder *Etropus crossotus*

Smallmouth flounder - *Etropus microstomus*
(Gill, 1864)

KEY FEATURES: Eyes on left side and small, separated by ridge; lateral line nearly straight. **COLOR:** Ocular side typically pale brown, with or without dusky blotches along lateral line; dark pigment between eyes and around mouth; often with diffuse dusky spots in a row near bases of dorsal and anal fins. **SIZE:** The maximum adult size is 15 cm (6 in) TL. **RANGE:** Found coastally from New York to Florida but rare south of North Carolina.

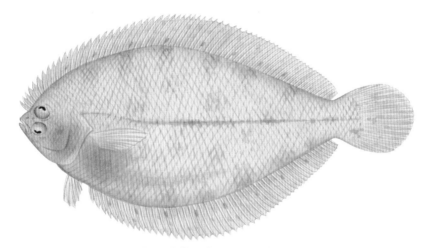

Smallmouth flounder *Etropus microstomus*

HABITAT AND HABITS: Smallmouth flounder are distributed from nearshore to about 37 m (120 ft) in depth; however, this species has been recorded as deep as 91 m (300 ft). Smallmouth flounder have a wide salinity tolerance. **OCCURRENCE IN THE CHESAPEAKE BAY:** Smallmouth flounder are common year-round residents of the lower Chesapeake Bay that frequent channels and mud bottoms.

REPRODUCTION: Spawning occurs in nearshore waters from June through October north of Cape Hatteras. Sexual maturity is reached at about 5 cm (2 in).

FOOD HABITS: Smallmouth flounder eat bottom-dwelling organisms including small fishes, worms, and shrimps.

IMPORTANCE: This species is of no commercial or recreational value because of its small size.

Summer flounder - *Paralichthys dentatus*
(Linnaeus, 1766)

KEY FEATURES: Eyes on left side and moderately large; lateral line strongly curved anteriorly. **COLOR:** Color variable, depending on background, typically brownish on ocular side with triangular pattern of ocelli, with one on midline and two on body edges at caudal peduncle; spots on anterior half of body smaller and variable in number and placement. Spots may be obscure in larger specimens. **SIZE:** Maximum adult size about 95 cm (3 ft) TL. Individuals greater than 60 cm (2 ft) TL are likely to be females. **RANGE:** Nova Scotia to south Florida, with greatest abundance between Massachusetts and North Carolina.

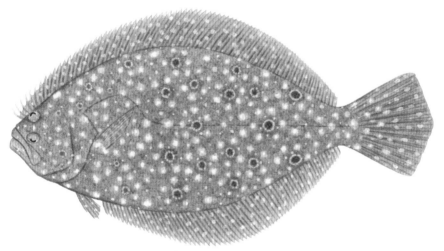

Summer flounder *Paralichthys dentatus*

HABITAT AND HABITS: Summer flounder typically congregate in shallow coastal and estuarine waters from late spring to fall but then migrate offshore to depths of 70–155 m (230–500 ft) during winter. Seasonal migrations are brought about by changes in photoperiod and water temperature. **OCCURRENCE IN THE CHESAPEAKE BAY:** Most summer flounder are visitors to the Chesapeake Bay from spring to autumn and migrate offshore during the winter months. However, some overwinter in the bay. Summer flounder are more common in the lower than the upper bay, but they extend as far north as the Elk and Sassafras Rivers near the head of the bay. Larvae enter the bay from October through May. Juvenile summer flounder utilize eelgrass beds. Adults typically occur in deep channels, over ridges, or over sandbars.

REPRODUCTION: Spawning occurs during the offshore migration from late summer to midwinter. Summer flounder can live to 20 years of age, with females living longer and growing larger than males.

FOOD HABITS: Like other flounders, this species partially conceals itself with sand and feeds on unsuspecting prey that venture too close. The diet of summer flounder consists primarily of shrimps, fishes, and squids.

IMPORTANCE: Summer flounder are of major commercial and recreational importance in the Chesapeake Bay region. From 1999 to 2008, Chesapeake Bay waters averaged 1,318 mt (2.9 million lb) per year commercially. For the period 2000–2009, the combined recreational harvest from bay waters of Maryland and Virginia averaged almost 650,000 fish per year, with about 85% of that number coming from Virginia's bay waters. Both commercial and recreational fisheries are presently managed with a scientifically based quota system.

Southern flounder - *Paralichthys lethostigma*
Jordan and Gilbert, 1884

KEY FEATURES: Eyes on left side, widely set; lateral line strongly curved anteriorly. COLOR: Body without prominent ocellated spots; ocular side with shades of olive brown, including inconspicuous dusky spots and blotches that are frequently absent in large individuals; blind side white or dusky. SIZE: Maximum adult size about 76 cm (30 in) TL. Males tend not to get larger than 40 cm (16 in) TL, whereas females are common to 50 cm (20 in) TL. RANGE: Known from the Chesapeake Bay to northern Mexico but absent from southern Florida.

HABITAT AND HABITS: Tolerating a wide salinity range, southern flounder are frequently found in brackish and freshwater habitats. Juveniles and adults are typically most abundant in shallow areas with aquatic vegetation on a muddy bottom. OCCURRENCE IN THE CHESAPEAKE BAY: Southern flounder are infrequent visitors to the lower Chesapeake Bay during late summer.

REPRODUCTION: Spawning occurs offshore from autumn to winter.

FOOD HABITS: Like other flounders, southern flounder partially bury themselves in the sand to ambush prey. While they will eat crustaceans such as shrimps and crabs, they primarily feed on small fishes.

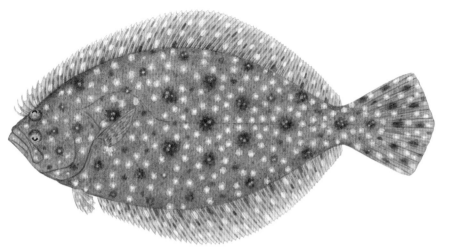

Southern flounder *Paralichthys lethostigma*

IMPORTANCE: Highly prized recreationally and commercially where common (in waters more southern than the Chesapeake Bay). Southern flounder are collected as sport fish by nighttime gigging in tidal creeks and salt marshes and by drift-fishing in channels.

Fourspot flounder - *Paralichthys oblongus*
(Mitchill, 1815)

KEY FEATURES: Eyes on left side, large and closely set; lateral line strongly curved anteriorly. **COLOR:** Ocular side mottled brownish gray with four conspicuous ocellated spots forming trapezoidal pattern, typically one pair near midbody and another pair near caudal peduncle; blind side of caudal, dorsal, and anal fins blackish, especially toward fin tips. **SIZE:** Maximum adult size to about 40 cm (1.3 ft) TL. **RANGE:** Known from Georges Bank to the Dry Tortugas, Florida. This species appears to be most abundant between New England and Delaware.

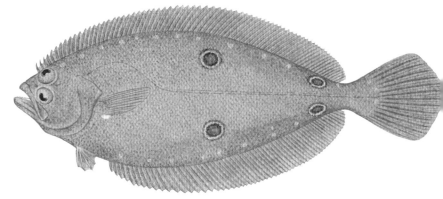

Fourspot flounder   *Paralichthys oblongus*

HABITAT AND HABITS: A cold-water species typically found in waters that are 14°C (57°F) or cooler. While this species has an extensive north-south distribution, it is common inshore only in the northern portion of its range. South of New York, fourspot flounder typically occur at depths exceeding 30 m (100 ft) on sand and mud bottoms. **OCCURRENCE IN THE CHESAPEAKE BAY:** Fourspot flounder are rare visitors to the lower Chesapeake Bay.

REPRODUCTION: Spawning apparently occurs offshore near the mid-shelf from late spring through early fall.

FOOD HABITS: Fourspot flounder feed on worms, shrimps, and other small crustaceans, with larger adults preying on crabs, squids, and small fishes.

IMPORTANCE: Fourspot flounder are of little commercial or recreational importance in the Chesapeake Bay region because of their small size and rarity. However, this species is taken in shelf waters by the winter trawl fishery, at least in small numbers.

Winter flounder (blackback flounder) -
Pseudopleuronectes americanus (Walbaum, 1792)

KEY FEATURES: Eyes located on right side of head; lateral line nearly straight, scarcely arched anteriorly. **COLOR:** Olive green with reddish-brown spots on ocular side, blind side white; background color may be pale or dusky; spots may be prominent, obscure, or almost absent. **SIZE:** Maximum adult size 64 cm (2 ft) TL. Inshore specimens, especially from the Chesapeake Bay, are in the size range of 20–40 cm (8–16 in) TL. **RANGE:** Occur coastally from Labrador to Georgia and most common between the Gulf of St. Lawrence and Delaware.

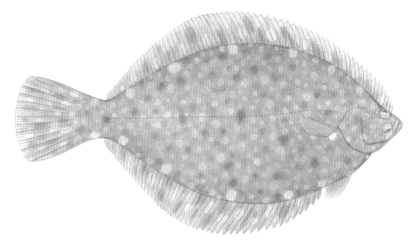

Winter flounder *Pseudopleuronectes americanus*

HABITAT AND HABITS: One of the most common flounders in North Atlantic coastal waters. Winter flounder are found most frequently on muddy or vegetated bottoms. This species migrates to shallow coastal waters in the fall and returns to deeper waters in the spring. **OCCURRENCE IN THE CHESAPEAKE BAY:** Winter flounder are found throughout the Chesapeake Bay, from the Susquehanna River to the bay mouth, primarily in winter. They are uncommon during the summer months, when they retreat to deeper water or migrate offshore. Winter flounder are a cold-water species that is much less common than in earlier times. Recent climate change and increasing bay water temperatures may eventually lead to extirpation of this species from the Chesapeake Bay.

REPRODUCTION: Spawning occurs in nearshore and estuarine waters from late winter to early spring.

FOOD HABITS: Winter flounder have a small mouth and thus are limited to feeding on smaller food items, such as small crustaceans and worms.

IMPORTANCE: Winter flounder were formerly valuable food fish in the Chesapeake Bay. They are enthusiastically pursued by recreational fishers using bloodworm bait while bottom-fishing during spawning runs in late winter and early spring.

Windowpane - *Scophthalmus aquosus* (Mitchill, 1815)

KEY FEATURES: Eyes located on left side of head; mouth large and nearly vertical; lateral line prominently arched anteriorly, lateral line visible on both sides of windowpanes. **COLOR:** Ocular side pale brown, mottled with numerous small brown and black spots of various sizes that continue onto fins; blind side white. **SIZE:** Maximum adult size about 46 cm (18 in) TL. **RANGE:** Known from the Gulf of St. Lawrence to Florida.

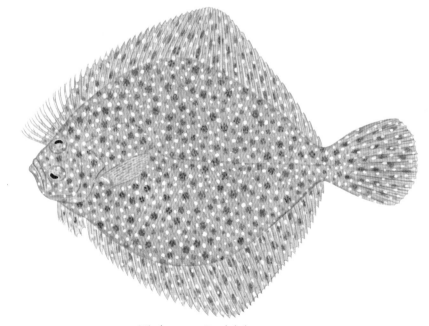

Windowpane *Scophthalmus aquosus*

HABITAT AND HABITS: The common name "windowpane" refers to the thinness of the body, which is almost transparent in some places. Windowpanes are shallow-water left-eyed flounders that are common in bays and estuaries along the East Coast of the United States. They are typically found on sand, silty sand, or mud bottoms at depths of a few meters to as great as 200 m (660 ft) but are most abundant at depths less than 50 m (165 ft). **OCCURRENCE IN THE CHESAPEAKE BAY:** Windowpanes are year-round Chesapeake Bay residents that are occasional to common in the middle bay, extending as far north as the Choptank River, and common to abundant in the lower bay.

REPRODUCTION: Spawning occurs throughout windowpanes' range from early spring to autumn on the inner continental shelf. In the Chesapeake Bay, spawning also occurs from spring to autumn, with a possible hiatus during summer months.

FOOD HABITS: Windowpane flounder are reported to feed on small fishes, shrimps, and other crustaceans. As windowpanes mature, they become more piscivorous.

IMPORTANCE: Windowpanes are of no commercial or recreational value in the Chesapeake Bay region.

Key to the Orders and Families
of Chesapeake Bay Fishes

Most of the fish families and orders treated in this key are morphologically diverse
and distributed worldwide. In many cases, the universal defining traits involve internal characteristics that require considerable knowledge and specimen preparation to
evaluate. To make the key accessible to persons who have limited familiarity with fish
anatomy or who lack the tools to use internal characteristics, we have limited the key
characteristics to those that require minimal use of magnification and no dissection.
The user should be aware that, by so limiting the characteristics, we have necessarily
restricted the utility of the key to the species that inhabit the Chesapeake Bay. Refer
to appendix 3 for the key to the species.

1a. Jaws absent, mouth in form of roundish, funnel-like opening with horny, conical
teeth; paired fins (pectoral and pelvic fins) absent; 7 pairs of gill openings present ....**Lampreys - Order Petromyzontiformes**
(Family Petromyzontidae) (p. 22)
1b. Jaws present; mouth not as above; paired fins present (1 or 2 pairs); fewer than
7 pairs of gill openings present 2

2a. Single pair of gill openings present...**(bony fishes) 9**
2b. Five pairs of gill openings present .... **(rays, sharks, and skates) 3**

3a. Gill openings visible from dorsal view and located laterally behind head; body
cross-section more or less rounded.. 4
3b. Gill openings not visible from dorsal view and located ventrally or ventrolaterally behind head; head and body dorsoventrally flattened, at least in front....... 6

4a. Anal fin absent; 1 stout anterior spine present on each dorsal fin ....
...................**Dogfish sharks - Order Squaliformes (Family Squalidae)** (p. 23)
4b. Anal fin present; spines absent on dorsal fins 5

5a. Eye with nictitating membrane ....
..... **Hammerhead, Hound, and Requiem sharks - Order Carcharhiniformes**
(Families Sphyrnidae, Triakidae, and Carcharhinidae) (pp. 303, 33, 304)

Eye of a sandbar shark showing nictitating membrane

5b. Eye without nictitating membrane ....
.... **Basking shark and Sand tigers - Order Lamniformes**
(Families Cetorhinidae and Odontaspididae) (pp. 34, 36)

6a. Snout in form of flat blade with lateral teeth ....
.... **Sawfishes - Order Pristiformes (Family Pristidae)** (p. 40)

6b. Snout not in form of flat blade with lateral teeth **7**

7a. Gill openings ventrolaterally on head; front margin of pectoral fins not connected to head; mouth terminal.................. **Angel sharks - Order Squatiniformes**
(Family Squatinidae) (p. 38)

7b. Gill openings ventrally on head; front margin of pectoral fins connected to head; mouth ventral **8**

8a. Dorsal fin absent or single dorsal fin present; caudal fin absent; elongate serrated spine present on tail (except in smooth butterfly ray)...............................
.... **Butterfly rays, Cownose rays, Eagle rays, and Whiptail stingrays -**
Order Myliobatiformes (Families Gymnuridae, Rhinopteridae,
Myliobatidae, and Dasyatidae) (pp. 305, 54, 305)

8b. Two dorsal fins present; caudal fin usually present (may be obscure in some large skates); elongate serrated spine absent on tail...
....**Skates - Order Rajiformes (Family Rajidae)** (p. 304)

9a. Caudal fin heterocercal or abbreviate heterocercal **10**

Strongly heterocercal tail of a sturgeon   Abbreviate heterocercal tail of a gar

9b. Caudal fin absent or not heterocercal **12**

10a. Caudal fin strongly heterocercal; 5 longitudinal rows of bony scutes present on body; snout projecting, with 4 ventral barbels; mouth ventral, with teeth absent or poorly developed................................**Sturgeons - Order Acipenseriformes**
(Family Acipenseridae) (p. 306)

10b. Caudal fin abbreviate heterocercal; scutes absent; snout not projecting; barbels absent; mouth terminal, with teeth well developed **11**

11a. Jaws very elongate, with many small needlelike teeth; body covered with hard diamond-shaped scales; dorsal fin short, located near caudal fin .... **Gars - Order Lepisosteiformes (Family Lepisosteidae)** (p. 64)
11b. Jaws not elongate; body with cycloid scales; dorsal fin long, extending more than half body length ....**Bowfins - Order Amiiformes (Family Amiidae)** (p. 66)

12a. Both eyes on same side of head... ....**American soles, Righteye flounders, Sand flounders, Tonguefishes, and Turbots - Order Pleuronectiformes (Families Achiridae, Pleuronectidae, Paralichthyidae, Cynoglossidae, and Scophthalmidae)** (pp. 276, 324, 278, 288)
12b. Eyes bilateral.. **13**

13a. Body eel-like... **14**
13b. Body not eel-like **15**

14a. Pelvic fin absent ..... **Conger eels and Freshwater eels - Order Anguilliformes (Families Congridae and Anguillidae)** (pp. 71, 70)
14b. Pelvic fin present, far forward, and barbel-like in appearance........................... ....**Cusk-eels - Order Ophidiiformes (Family Ophidiidae)** (p. 109)

15a. Adipose fin present **16**
15b. Adipose fin absent..**17**

16a. Barbels present on head; spines present in dorsal and pectoral fins ....**North American catfishes and Sea catfishes - Order Siluriformes (Families Ictaluridae and Ariidae)** (pp. 308, 96)
16b. Barbels absent; spines absent in fins ....**Lizardfishes - Order Aulopiformes (Family Synodontidae)** (p. 108)

17a. Dorsal and/or anal fin with 1 or more spines.. **18**
17b. Dorsal and anal fins without apparent spines **26**

18a. Pelvic fin absent or represented by midventral spine or tubercle **19**
18b. Pelvic fin present **20**

19a. Gill opening in form of small vertical slit anterior to and approximately as long as pectoral-fin base................ **Boxfishes, Filefishes, Porcupinefishes, Puffers, and Triggerfishes - Order Tetraodontiformes (Families Ostraciidae, Monacanthidae, Diodontidae, Tetraodontidae, and Balistidae)** (pp. 273, 323, 268, 323, 267)

19b. Gill opening typically much longer than pectoral-fin base ....
....**Perch-like fishes, in part - Order Perciformes**
(go to appendix 2, Key to the Families of Perciformes Fishes)

20a. Pelvic fin jugular in position, well in advance of pectoral fin **21**
20b. Pelvic fins thoracic or abdominal in position, under or posterior to pectoral fin
.... **22**

21a. Anteriormost dorsal spine free, located on head, and tipped with esca (angling
bait); gill opening in form of small hole located in pectoral-fin axil ....
......................**Goosefishes - Order Lophiiformes (Family Lophiidae)** (p. 118)

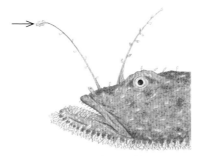

Esca of a goosefish

21b. Dorsal spines stout and short, located on body and not formed into angling
device; gill opening located anterior to pectoral-fin base; pectoral-fin axil with
small opening into blind pouch .... **Toadfishes - Order Batrachoidiformes**
(Family Batrachoididae) (p. 116)

22a. Spiny stay present on cheek below eye (extension of third infraorbital bone);
pectoral fin greatly enlarged (except in lumpfishes) ....
....**Sculpins and Searobins - Order Scorpaeniformes**
(Families Cottidae and Triglidae) (pp. 148, 312)
22b. Spiny stay absent on cheek below eye; pectoral fin typically not greatly en-
larged.. **23**

23a. First dorsal fin composed of 2–6 stout unconnected spines remote from second
dorsal fin; pelvic fin subabdominal, with 1 stout spine and 0–2 rays..................
....**Sticklebacks - Order Gasterosteiformes (Family Gasterosteidae)** (p. 312)
23b. Dorsal spines connected by membrane or, if unconnected, contiguous with remaining dorsal elements; pelvic fin abdominal to thoracic, usually with 1 spine
and 5 rays (combtooth blennies - Perciformes: Blenniidae are a notable excep-
tion)... **24**

24a. Pectoral fin high on body, with base mostly above lateral midline; pelvic fin
abdominal; lateral line absent **25**

24b. Pectoral fin low on body, with base mostly or entirely below lateral midline; pelvic fin abdominal or thoracic; lateral line variously developed but present....
....**Perch-like fishes, in part - Order Perciformes**
(go to appendix 2, Key to the Families of Perciformes Fishes)

25a. Adipose eyelid present ....**Mullets - Order Mugiliformes**
(Family Mugilidae) (p. 309)

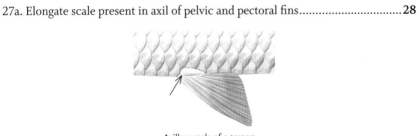

Adipose eyelid of a mullet

25b. Adipose eyelid absent.............**New World silversides - Order Atheriniformes**
(Family Atherinopsidae) (p. 310)

26a. Mouth at end of long, tubelike snout; body encased in series of bony rings; pelvic fin absent...**Pipefishes - Order Gasterosteiformes**
(Family Syngnathidae) (p. 312)

26b. Mouth not at end of elongate snout; body not encased in bony rings; pelvic fin present..**27**

27a. Elongate scale present in axil of pelvic and pectoral fins...............................**28**

Axillary scale of a tarpon

27b. Elongate scale absent in axil of pelvic and pectoral fins **31**

28a. Bony gular plate present on underside of head, between arms of lower jaw.......
....**Tarpons and Tenpounders - Order Elopiformes**
(Families Megalopidae and Elopidae) (pp. 68, 67)

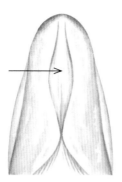

Gular plate of a tarpon

28b. Bony gular plate absent or not externally visible 29

29a. Belly with sharp-angled keel formed by median row of scutes...........................
.........................**Herrings - Order Clupeiformes (Family Clupeidae)** (p. 306)
29b. Belly rounded, without median row of scutes 30

30a. Mouth terminal or superior, snout not overhanging...
..............**Round herrings - Order Clupeiformes (Family Clupeidae)** (p. 306)
30b. Mouth inferior, with snout overhanging ....
....................**Anchovies - Order Clupeiformes (Family Engraulidae)** (p. 307)

31a. Belly with sucking disk completely formed from pelvic fins; body globose and
triangular in cross-section; skin with bony tubercles or dermal warts...............
.........**Lumpfishes - Order Scorpaeniformes (Family Cyclopteridae)** (p. 148)
31b. Belly without sucking disk; body not globose or triangular in cross-section; bony
tubercles or dermal warts on skin absent..**32**

32a. Pelvic fin jugular or thoracic in position, typically more anterior than pectoral
fin ....**Cods, Merlucciid hakes, and Phycid hakes - Order Gadiformes
(Families Gadidae, Merlucciidae, and Phycidae)** (pp. 309, 113, 309)
32b. Pelvic fin abdominal in position, well posterior to pectoral fin **33**

33a. Lower jaw or both jaws greatly elongated, to form beak................................**34**
33b. Neither jaw elongated to form beak **36**

34a. Both jaws elongated to form beak...**35**
34b. Lower jaw considerably elongated but upper jaw normal ....
..............**Halfbeaks - Order Beloniformes (Family Hemiramphidae)** (p. 311)

35a. Snout in form of ducklike beak; dorsal and anal fins short and rounded; pectoral
fin low on body ....**Pikes - Order Esociformes (Family Esocidae)** (p. 309)
35b. Snout in form of needlelike beak; dorsal and anal fins not short and rounded;
pectoral fin high on body..........................**Needlefishes - Order Beloniformes
(Family Belonidae)** (p. 310)

36a. Scales absent on head and cheeks ....
.... **Carps, Minnows, and Suckers - Order Cypriniformes**
(Families Cyprinidae and Catostomidae) (p. 307)
36b. Scales present on head and cheeks.. **37**

37a. Mouth protrusible........................ **Livebearers, Pupfishes, and Topminnows -**
Order Cyprinodontiformes (Families Poeciliidae, Cyprinodontidae,
and Fundulidae) (pp. 140, 130, 311)
37b. Mouth not protrusible...
.... **Mudminnows - Order Esociformes (Family Umbridae)** (p. 107)

APPENDIX 2. Key to the Families of Perciformes Fishes in the Chesapeake Bay: Perch-like fishes - Order Perciformes

This extremely diverse order comprises more than 150 families, nearly 1,300 genera, and possibly 13,000 species. About three-quarters of all Perciformes species are marine shorefishes. Twenty-eight families of perciform fishes are represented in bay waters and more than half of the species treated in this book belong to this order.

1a. Pelvic fins absent.. 2
1b. Pelvic fins present 4

2a. Body short and very deep .... **Butterfishes - Family Stromateidae** (p. 313)
2b. Body elongate and not deep... 3

3a. Caudal fin absent; body ribbon-like ....
.................................**Cutlassfishes - Family Trichiuridae** (p. 155)
3b. Caudal fin present; body not ribbon-like ....
.................................**Sand lances - Family Ammodytidae** (p. 156)

4a. Anal fin preceded by two spines detached from main body of fin.....................
.................................**Jacks, in part - Family Carangidae** (p. 313)

Anal-fin spines of some jacks

4b. Anal-fin spines contiguous with main body of fin 5

5a. Three to nine free spines present in dorsal fin (may be connected by membrane
in Carangidae less than 30 mm, or 1.2 in, SL)................................. 6
5b. Free spines absent in dorsal fin.. 7

6a. Head and body strongly compressed; color predominantly silver......................
.................................**Jacks, in part - Family Carangidae** (p. 313)
6b. Head depressed; body only slightly compressed; color not predominantly silver
.................................**Cobia - Family Rachycentridae** (p. 180)

7a. Dorsal spines absent............................**Remoras - Family Echeneidae** (p. 315)
7b. Dorsal spines present 8

8a. Dorsal fin divided, or nearly divided, into distinct spinous and soft dorsal portions..**9**

8b. Spinous and soft portions of dorsal fin continuous **19**

9a. Spinous and soft dorsal fins widely separated **10**

9b. Spinous and soft dorsal fins contiguous (connected basally by membrane).....**11**

10a. Dorsal and anal fins followed by series of small, detached finlets.......................
....**Mackerels, in part - Family Scombridae** (p. 315)

10b. Dorsal and anal fins not followed by series of small, detached finlets................
....**Barracudas - Family Sphyraenidae** (p. 316)

11a. Dorsal and anal fins followed by series of small, detached finlets......................
....**Mackerels, in part - Family Scombridae** (p. 315)

11b. Dorsal and anal fins not followed by series of small, detached finlets............**12**

12a. Eye located dorsally; mouth oriented nearly vertically, with fringed lips............
....**Stargazers - Family Uranoscopidae** (p. 191)

12b. Eye more or less lateral; mouth not nearly vertical; lips without fringe..........**13**

13a. Lateral line (and scales) extending onto central caudal fin rays to posterior margin ....**Drums and croakers - Family Sciaenidae** (p. 317)

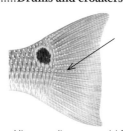

Lateral line extending onto caudal fin

13b. Lateral line not extending onto caudal fin **14**

14a. Dorsal and anal soft fins covered with fine scales...**15**

14b. Dorsal and anal soft fins not covered with fine scales **16**

15a. Body very deep and compressed; teeth small, in bands.......................................
....**Spadefishes - Family Ephippidae** (p. 213)

15b. Body elongate, moderately compressed; outer row of teeth large and blade-like
....**Bluefish - Family Pomatomidae** (p. 214)

16a. Pelvic fins formed into sucking disk **17**

United pelvic fins of a goby

16b. Pelvic fins not formed into sucking disk **18**

17a. Head dorsoventrally flattened........ **Clingfishes - Family Gobiesocidae** (p. 216)
17b. Head not dorsoventrally flattened .... **Gobies - Family Gobiidae** (p. 318)

18a. Three anal-fin spines present . **Temperate basses - Family Moronidae** (p. 319)
18b. One or two anal-fin spines present .... **Perches - Family Percidae** (p. 319)

19a. Spines in dorsal and anal fins flexible **20**
19b. Spines in dorsal and anal fins stiff **21**

20a. Teeth close-set and comb-like; ocular cirri usually present................................
.... **Combtooth blennies - Family Blenniidae** (p. 319)

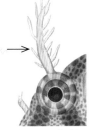

Ocular cirrus of a blenny

20b. Teeth prominent and sharp; no ocular cirri present ....
.... **Snakeheads - Family Channidae** (p. 232)

21a. Pelvic axillary process present **22**

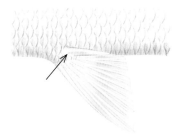

Pelvic axillary process of an Irish pompano

21b. Pelvic axillary process absent **26**

22a. Caudal fin rounded ....**Butterflyfishes - Family Chaetodontidae** (p. 234)
22b. Caudal fin forked **23**

23a. Mouth greatly protrusible.................... **Mojarras - Family Gerreidae** (p. 320)

Protrusible mouth of a mojarra

23b. Mouth moderately protrusible **24**

24a. Posterior jaw teeth molar-like ....**Porgies - Family Sparidae** (p. 320)

Molar-like teeth of a sheepshead

24b. Molar-like teeth absent **25**

25a. Upper jaw with 1 or 2 pairs of greatly enlarged caninelike teeth ....
....**Snappers - Family Lutjanidae** (p. 321)
25b. Upper jaw without enlarged caninelike teeth ....
....**Grunts - Family Haemulidae** (p. 321)

26a. Last (innermost) pelvic ray connected to body by membrane...........................
.... **Sunfishes - Family Centrarchidae** (p. 321)
26b. Last (innermost) pelvic ray not connected to body by membrane.................**27**

27a. Jaw teeth projecting; no slit-like opening into pharynx posterior to last (fourth)
gill arch....................................... **Wrasses - Family Labridae** (p. 322)

Projecting teeth of a tautog

27b. Jaw teeth not projecting; slit-like opening into pharynx present posterior to last (fourth) gill arch **28**

28a. Caudal fin forked ....**Medusafishes - Family Centrolophidae** (p. 263)
28b. Caudal fin rounded, truncate, or emarginate..
..................................**Sea basses and groupers - Family Serranidae** (p. 323)

After you have successfully identified the family to which your fish belongs (see appendices 1 and 2), proceed to the appropriate key to species of that family (below) and continue the process until your couplet selection yields the common and scientific name of a species. Note that when a family is represented by a single species in Chesapeake Bay waters, the common and scientific names of the species and the page number of the species account are provided in the family key, and no key to the species of the family is given.

If you have accurately compared the characters of your fish with those in the key couplets, your fish is correctly identified when you reach the species names. Your identification can be verified by referring to the more extensive description in the species account and to the illustrations(s) provided. For those species only found in appendix 4, Fish Species Rarely Recorded from the Chesapeake Bay, please consult another identification guide or bring your fish to the attention of someone who specializes in the study of Chesapeake Bay fishes.

An asterisk at the end of an entry means that the fish is treated in appendix 4, Fish Species Rarely Recorded from the Chesapeake Bay.

Key to the Species of Hammerhead Sharks (Family Sphyrnidae) in the Chesapeake Bay

1a. Head flattened dorsoventrally and shovel- or bonnet-shaped, anterior contour of head evenly rounded at midline (see figs. p. 32) .... **bonnethead** - *Sphyrna tiburo* (p. 32)

1b. Head broad and T-shaped, anterior contour of head moderately convex (see figs. p. 31)... 2

2a. Median indentation present on anterior margin of snout; free rear tip of second dorsal fin nearly reaching origin of upper caudal fin; base of anal fin noticeably longer than base of second dorsal fin (see figs. p. 31, *top*)..................................... ....**scalloped hammerhead** - *Sphyrna lewini* (p. 30)

2b. Median indentation absent on anterior margin of snout; free rear tip of second dorsal fin termination well anterior to origin of upper caudal fin; base of anal fin about as long as base of second dorsal fin (see figs. p. 31, *bottom*) ....**smooth hammerhead** - *Sphyrna zygaena* (p. 30)

Key to the Species of Requiem Sharks (Family Carcharhinidae) in the Chesapeake Bay

1a. Base of second dorsal fin at least three-fourths as long as base of first dorsal fin, dorsal fins nearly equal in size.. **lemon shark - *Negaprion brevirostris*** (p. 326)*

1b. Base of second dorsal fin considerably shorter than base of first dorsal fin....... 2

2a. Spiracle present .... **tiger shark - *Galeocerdo cuvier*** (p. 326)*

2b. Spiracle absent... 3

3a. Origin of second dorsal fin well dorsoposterior to origin of anal fin; posterior margin of anal fin straight or weakly concave ....
.... **Atlantic sharpnose shark - *Rhizoprionodon terraenovae*** (p. 28)

3b. Origin of second dorsal fin typically nearly perpendicular to origin of anal fin; posterior margin of anal fin deeply concave or deeply notched........................ 4

4a. Midline of back between dorsal fins with low but distinct ridge of skin 5

4b. Midline of back between dorsal fins smooth, without ridge of skin................. 6

5a. First dorsal fin very high and nearly triangular; origin of first dorsal fin far anterior, perpendicular to insertion of pectoral fin...
.... **sandbar shark - *Carcharhinus plumbeus*** (p. 26)

5b. First dorsal fin moderate in height, with broadly arched anterior margin; origin of first dorsal fin perpendicular to free rear tips of pectoral fin ....
.... **dusky shark - *Carcharhinus obscurus*** (p. 25)

6a. Snout short and broadly rounded; fins without black tips...................................
.... **bull shark - *Carcharhinus leucas*** (p. 24)

6b. Snout long and pointed; fins with black tips...
....**blacktip shark - *Carcharhinus limbatus*** (p. 326)*

Key to the Species of Skates (Family Rajidae) in the Chesapeake Bay

1a. Large thorny scales absent along middorsal zone of disk; ventral surface of disk marked with dusky dots and dashes.....**barndoor skate - *Dipturus laevis*** (p. 00)

1b. One or more rows of thorny scales present along middorsal zone of disk; ventral surface of disk not marked with dusky dots and dashes.................................... 2

2a. Snout acutely pointed; dorsal surface of disk with distinctive dusky bars...........
.... **clearnose skate - *Raja eglanteria*** (p. 00)

2b. Snout bluntly rounded or only slightly pointed; dorsal surface of disk with small dusky spots... 3

3a. Several large ocellar spots typically present on dorsal surface of disk; 63 or more series of teeth present in upper jaw .... **winter skate - *Leucoraja ocellata*** (p. 58)

3b. Ocellar spots typically lacking on dorsal surface of disk; 53 or fewer series of teeth in upper jaw .... **little skate -** *Leucoraja erinacea* (p. 58)

Key to the Species of Butterfly Rays (Family Gymnuridae) in the Chesapeake Bay

1a. Typically 1 or 2 serrated spines present on tail and long tentacle present at each spiracle .... **spiny butterfly ray -** *Gymnura altavela* (p. 50)
1b. Tail spine and spiracular tentacles always absent ....
.... **smooth butterfly ray -** *Gymnura micrura* (p. 51)

Key to the Species of Whiptail Stingrays (Family Dasyatidae) in the Chesapeake Bay

1a. Lateral corners of disk wings narrowly rounded or abruptly angled 2
1b. Lateral corners of disk wings broadly and evenly rounded 3

2a. Finlike fold of skin along underside of tail about as wide as height of tail; dorsal surface of tail with single ridge or keel posterior to spine; sides of tail without thorns .... **southern stingray -** *Dasyatis americana* (p. 42)
2b. Finlike fold of skin along underside of tail only about half as wide as height of tail; dorsal surface of tail without ridge or keel posterior to spine; in large individuals, sides of tail with thorns ....
.... **roughtail stingray -** *Dasyatis centroura* (p. 44)

3a. Distance from eye to tip of snout considerably longer than distance between spiracles; front outline of disk wings concave near tip of snout.........................
.... **Atlantic stingray -** *Dasyatis sabina* (p. 46)
3b. Distance from eye to tip of snout shorter than distance between spiracles; front outline of disk wings weakly convex near tip of snout.......................................
.... **bluntnose stingray -** *Dasyatis say* (p. 48)

Key to the Species of Eagle Rays (Family Myliobatidae) in the Chesapeake Bay

1a. One series of large teeth present in each jaw ....
.... **spotted eagle ray -** *Aetobatus narinari* (p. 327)*
1b. More than 1 (typically 7–9) series of teeth present in each jaw ....
.... **bullnose ray -** *Myliobatis freminvillei* (p. 52)

Key to the Species of Sturgeons (Family Acipenseridae) in the Chesapeake Bay

The head shape and relative snout length change drastically with age in sturgeons. Because of allometric growth, the snout becomes relatively shorter and blunter. Consequently, key characters based on snout length as a proportion of total length are of little use and have frequently led to misidentifications. In particular, Atlantic sturgeon (*Acipenser oxyrinchus*) of moderate size (approximately 1 m, or 3.3 ft, TL) may be misidentified as shortnose sturgeon (*A. brevirostrum*).

1a. Mouth small, with width inside lips usually less than 62% (43–66%) of interorbital width; bony scutes present between anal fin and midlateral scutes ....
.... **Atlantic sturgeon** - *Acipenser oxyrinchus* (p. 63)
1b. Mouth large, with width inside lips greater than 62% (63–81%) of interorbital width; bony scutes absent between anal fin and midlateral scutes ....
.... **shortnose sturgeon** - *Acipenser brevirostrum* (p. 62)

Key to the Species of Herrings (Family Clupeidae) in the Chesapeake Bay

1a. Abdominal scutes absent; 1 pelvic scute present...
....**round herring** - *Etrumeus teres* (p. 82)
1b. Abdominal scutes and 1 pelvic scute present 2

2a. Last ray of dorsal fin long and filamentous...3
2b. Last ray of dorsal fin not long and filamentous 5

3a. Predorsal midline without scales; gill rakers numbering more than 180 4
3b. Predorsal midline with scales; gill rakers numbering fewer than 120................
.....................................**Atlantic thread herring** - *Opisthonema oglinum* (p. 83)

4a. Snout bulbous and fleshy; upper jaw slightly projecting; more than 50 scales in lateral series; anal fin typically with 29–35 rays; prepelvic scutes typically numbering 18 or more........................ **gizzard shad** - *Dorosoma cepedianum* (p. 80)
4b. Snout more pointed; upper jaw not projecting; fewer than 50 scales in lateral series; anal fin typically with 20–25 rays; prepelvic scutes typically numbering fewer than 18 .... **threadfin shad** - *Dorosoma petenense* (p. 80)

5a. Predorsal scales on either side of midline enlarged; pelvic fin with 7 elements..
....**Atlantic menhaden** - *Brevoortia tyrannus* (p. 77)
5b. Predorsal scales on either side of midline not enlarged; pelvic fin typically with 9 elements 6

6a. Body depth 20–26% SL; dark spot absent just behind opercle............................
....**Atlantic herring** - *Clupea harengus* (p. 79)
6b. Body depth 30–37% SL; dark spot present just behind opercle, sometimes followed by series of smaller spots...7

7a. Eye diameter usually less than greatest cheek depth **8**
7b. Eye diameter usually equal to or greater than greatest cheek depth **9**

8a. Gill rakers on lower limb of first arch numbering 59–76; teeth absent in fish greater than 15 cm (6 in) SL; tongue with pigment in medial area; peritoneum pale to silvery .... **American shad - *Alosa sapidissima*** (p. 76)
8b. Gill rakers on lower limb of first arch numbering 18–23; minute teeth present in fish at all sizes; tongue without pigment in medial area; peritoneum grayish, peppered with black .... **hickory shad - *Alosa mediocris*** (p. 74)

9a. Eye diameter greater than snout length; dorsum grayish green; peritoneum pale with dusky dots .... **alewife - *Alosa pseudoharengus*** (p. 72)
9b. Eye diameter less than or equal to snout length; dorsum deep bluish green; peritoneum black to dusky .... **blueback herring - *Alosa aestivalis*** (p. 72)

Key to the Species of Anchovies (Family Engraulidae) in the Chesapeake Bay

1a. Origin of anal fin perpendicular to anterior portion or midportion of dorsal fin; anal fin with 23–31 rays .... **bay anchovy - *Anchoa mitchilli*** (p. 84)
1b. Origin of anal fin perpendicular to posterior portion of dorsal fin; anal fin with 18–24 rays..................................... **striped anchovy - *Anchoa hepsetus*** (p. 84)

Key to the Species of Suckers (Family Catostomidae) in the Chesapeake Bay

1a. Dorsal fin with more than 23 rays .... **quillback - *Carpiodes cyprinus*** (p. 85)
1b. Dorsal fin with fewer than 16 rays **2**

2a. More than 50 scales present in lateral line ....
.... **white sucker - *Catostomus commersonii*** (p. 86)
2b. Fewer than 50 scales present in lateral line ....
...............................**shorthead redhorse - *Moxostoma macrolepidotum*** (p. 86)

Key to the Species of Carps and Minnows (Family Cyprinidae) in the Chesapeake Bay

1a. First ray of dorsal and anal fins hardened and spinelike; dorsal fin with more than 12 rays..**2**
1b. First ray of dorsal and anal fins not hardened and spine-like; dorsal fin with fewer than 12 rays **3**

2a. Two barbels present on each side of upper jaw ....
....**common carp - *Cyprinus carpio*** (p. 90)
2b. Barbels absent on jaws .... **goldfish - *Carassius auratus*** (p. 88)

3a. Anal fin with 10 or more rays; ventral midline between pelvic and anal fins form-
     ing keel ....**golden shiner** - *Notemigonus crysoleucas* (p. 93)
3b. Anal fin typically with 10 or fewer rays; ventral midline between pelvic and anal
     fins not forming keel **4**

4a. Peritoneum black and visible through belly; intestine very long and coiled (vis-
     ible upon dissection)...... **eastern silvery minnow** - *Hybognathus regius* (p. 92)
4b. Peritoneum typically silvery or speckled, sometimes visible through belly; intes-
     tine not very long and coiled (sometimes visible without dissection)............... **5**

5a. Posterior interradial membranes of dorsal fin with dusky pigment ....
     ....**spotfin shiner** - *Cyprinella spiloptera* (p. 94)
5b. Posterior interradial membranes of dorsal fin without dusky pigment **6**

6a. Black longitudinal stripe extending from opercle to caudal fin ....
     .... **spottail shiner** - *Notropis hudsonius* (p. 94)
6b. Black longitudinal stripe extending from snout to caudal fin ....
     ....**bridle shiner** - *Notropis bifrenatus* (p. 94)

Key to the Species of North American Catfishes (Family Ictaluridae) in the Chesapeake Bay

1a. Adipose fin not attached to caudal fin.. **2**
1b. Adipose fin attached to caudal fin...
     ....**margined madtom** - *Noturus insignis* (p. 102)

2a. Caudal fin distinctly forked **3**
2b. Caudal fin rounded, truncate, or slightly emarginate but not forked................ **5**

3a. Anal fin with fewer than 26 rays; tail moderately forked ....
     ....**white catfish** - *Ameiurus catus* (p. 97)
3b. Anal fin with typically 26 or more rays; tail deeply forked **4**

4a. Outer margin of anal fin rounded; anal fin with fewer than 30 rays; body typi-
     cally with dusky spots................. **channel catfish** - *Ictalurus punctatus* (p. 101)
4b. Outer margin of anal fin straight; anal fin with 30 or more rays; body lacking
     dusky spots... **blue catfish** - *Ictalurus furcatus* (p. 100)

5a. Lower jaw strongly projecting beyond upper jaw; snout greatly flattened ....
     ....**flathead catfish** - *Pylodictis olivaris* (p. 104)
5b. Lower jaw not projecting beyond upper jaw; snout not greatly flattened.......... **6**

6a. Chin barbels whitish; anal fin with typically 24–27 rays; caudal-fin margin
     straight or nearly so....................... **yellow bullhead** - *Ameiurus natalis* (p. 98)
6b. Chin barbels grayish or blackish; anal fin rays with typically fewer than 24 rays;
     caudal-fin margin slightly notched ....
     ....**brown bullhead** - *Ameiurus nebulosus* (p. 98)

Key to the Species of Pikes (Family Esocidae) in the Chesapeake Bay

1a. Branchiostegal rays numbering 16 or fewer, typically 12–14; snout length contained more than 8 times in TL; suborbital bar with slight ventroposterior slant; side with 20–36 wavy bars .... **redfin pickerel -** *Esox americanus* (p. 105)

1b. Branchiostegal rays numbering 14 or more; snout less than 7 times TL; suborbital bar vertical; sides with reticulated (chain-like) pattern............................
.... **chain pickerel -** *Esox niger* (p. 106)

Key to the Species of Cods (Family Gadidae) in the Chesapeake Bay

1a. Lower jaw projecting beyond upper jaw; chin barbel absent or tiny (present only in juvenile pollock) ....**pollock -** *Pollachius virens* (p. 112)

1b. Upper jaw projecting beyond lower jaw; chin barbel present and obvious..........
.... **Atlantic cod -** *Gadus morhua* (p. 110)

Key to the Species of Phycid Hakes (Family Phycidae) in the Chesapeake Bay

1a. First dorsal fin without elongate filament...
....**spotted hake -** *Urophycis regia* (p. 115)

1b. First dorsal fin with elongate filament **2**

2a. Lateral scale rows numbering 95–120; 3 gillrakers present on upper limb of first arch; scales between lateral line and first dorsal-fin base numbering about 7–10
....**red hake -** *Urophycis chuss* (p. 114)

2b. Lateral scale rows numbering about 120–150; 2 gillrakers on upper limb of first arch; scales between lateral line and first dorsal-fin base numbering about 11–13.. **white hake -** *Urophycis tenuis* (p. 328)*

Key to the Species of Mullets (Family Mugilidae) in the Chesapeake Bay

Note: Young mullets of 3.5–6.5 cm (1.4–2.5 in) SL are in transition to the juvenile stage and many have not developed the scalation or pigment patterns of juveniles and adults. At these sizes, both mullet species found in the Chesapeake Bay are simply little silvery fishes that may be reliably distinguished only by the number of anal-fin elements.

Key to Mullets Less Than 6.5 cm (2.5 in) SL

1a. Anal fin with total of 11 elements, either 2 spines and 9 rays or 3 spines and 8 rays .... **striped mullet -** *Mugil cephalus* (p. 120)

1b. Anal fin with total of 12 elements, either 2 spines and 10 rays or 3 spines and 9 rays .... **white mullet -** *Mugil curema* (p. 121)

Key to Mullets Greater Than 6.5 cm (2.5 in) SL

1a. Anal fin with 3 spines and 8 rays; second dorsal fin and anal fin with few or no
    scales; lateral scales with central dusky blotch forming longitudinal dusky stripe
    on each scale row .... **striped mullet** - *Mugil cephalus* (p. 120)
1b. Anal fin with 3 spines and 9 rays; second dorsal fin and anal fin densely scaled,
    nearly to margin; lateral scale rows without definite dusky stripes.....................
    .... **white mullet** - *Mugil curema* (p. 121)

Key to the Species of New World Silversides (Family Atherinopsidae) in the Chesapeake Bay

1a. Scales rough, with posterior margins fringed; bases of dorsal and anal fins with
    large deciduous scales .... **rough silverside** - *Membras martinica* (p. 122)
1b. Scales smooth, posterior margins without fringe; bases of dorsal and anal fins
    without scales 2

2a. Origin of spinous dorsal fin posterior to a vertical through anus; lateral scales
    numbering 39–55; segmented anal-fin rays numbering 19–29 (typically 23–
    25).. **Atlantic silverside** - *Menidia menidia* (p. 124)
2b. Origin of spinous dorsal fin anterior to a vertical through anus; lateral scales
    numbering 28–42; segmented anal-fin rays numbering 8–19 (typically 16) ......
    ....**inland silverside** - *Menidia beryllina* (p. 123)

Key to the Species of Needlefishes (Family Belonidae) in the Chesapeake Bay

1a. Body strongly compressed; side with series of vertical bars; pectoral fin falcate;
    anal fin with 24–28 rays......................**flat needlefish** - *Ablennes hians* (p. 125)
1b. Body roundish in cross-section; side without vertical bars; pectoral fin not fal-
    cate; anal fin with 13–24 rays 2

2a. Dorsal fin with 14–17 rays; lateral keel absent on caudal peduncle; caudal fin
    with concave margin, not deeply forked ....
    ....**Atlantic needlefish** - *Strongylura marina* (p. 126)
2b. Dorsal fin with 21–26 rays; lateral keel on caudal peduncle; caudal fin deeply
    forked 3

3a. Dorsal fin with 19–23 rays (typically 22 or 23); anal fin rays with 18–22 rays
    (typically 20 or 21)............................ **houndfish** - *Tylosurus crocodilus* (p. 127)
3b. Dorsal fin with 23–26 rays (typically 24); anal fin with 20–24 rays (typically 21
    or 22)... **Atlantic agujon** - *Tylosurus acus* (p. 328)*

Key to the Species of Halfbeaks (Family Hemiramphidae) in the Chesapeake Bay

1a. Dorsal-fin origin markedly more anterior than anal-fin origin; origin of pelvic fin closer to base of caudal fin than to opercle; caudal fin deeply forked, with ventral lobe almost twice as long as dorsal lobe ....**ballyhoo** - *Hemiramphus brasiliensis* (p. 128)

1b. Dorsal-fin origin opposite or only slightly more anterior than anal-fin origin; origin of pelvic fin about midway between opercle and base of caudal fin; caudal fin moderately forked, with lobes nearly equal in length.................................... .... **false silverstripe halfbeak** - *Hyporhamphus meeki* (p. 129)

Key to the Species of Topminnows (Family Fundulidae) in the Chesapeake Bay

1a. Teeth in single row; longitudinal scale rows numbering fewer than 30.............. .... **rainwater killifish** - *Lucania parva* (p. 138)

1b. Teeth in more than one row, with outer row sometimes large and inner row or rows sometimes small; longitudinal scale rows numbering more than 30 2

2a. Longitudinal scale rows numbering more than 40.. .... **banded killifish** - *Fundulus diaphanus* (p. 132)

2b. Longitudinal scale rows numbering 31–39 3

3a. Dorsal fin with 8 or 9 rays (usually 8); anal-fin base longer than dorsal-fin base; gill opening restricted, with dorsal end opposite or slightly dorsal to dorsalmost edge of pectoral-fin base....................**spotfin killifish** - *Fundulus luciae* (p. 134)

3b. Dorsal fin with 10–15 rays; anal-fin base shorter than dorsal-fin base; gill opening not restricted, dorsal end markedly dorsal to pectoral-fin base near postero-dorsal corner of opercle.. 4

4a. Snout long, pointed, and about twice as long as eye diameter; dorsal-fin with 13–15 rays; adult females with several irregular longitudinal stripes................. ....**striped killifish** - *Fundulus majalis* (p. 136)

4b. Snout short, rounded, and about equal in length to eye diameter; dorsal fin with 10–12 rays; adult females without longitudinal stripes but sometimes with vertical bars.. 5

5a. Anal-fin with 9–11 rays (usually 10); dorsal-fin origin perpendicular to anal-fin origin; both sexes usually with ocellus or 1 or 2 black blotches on last few dorsal-fin rays...**Bayou killifish** - *Fundulus pulvereus* (p. 137)

5b. Anal-fin with 10–12 rays (usually 11); dorsal-fin origin more anterior than anal-fin origin; females without ocellus or blotch on dorsal fin .... **mummichog** - *Fundulus heteroclitus* (p. 133)

Key to the Species of Sticklebacks (Family Gasterosteidae) in the Chesapeake Bay

1a. Vertical elongate bony scutes present along side; typically 3 dorsal spines present .... **threespine stickleback** - *Gasterosteus aculeatus* (p. 142)

1b. Bony scutes absent along side; typically 4 dorsal spines present........................
.... **fourspine stickleback** - *Apeltes quadracus* (p. 141)

Key to the Species of Pipefishes (Family Syngnathidae) in the Chesapeake Bay

Note: In pipefishes, the first trunk ring bears the pectoral fins, and the last trunk ring surrounds the anus.

1a. Tail prehensile; caudal fin absent ....
.... **lined seahorse** - *Hippocampus erectus* (p. 144)

1b. Tail not prehensile; caudal fin present.. 2

2a. Trunk rings numbering 16–19 (typically 17); rings totaling 46–58 (typically 48–53); dorsal fin covering 0.25–3.0 (typically less than 2.5) trunk rings; dorsal fin with typically fewer than 32 rays ....
.... **dusky pipefish** - *Syngnathus floridae* (p. 145)

2b. Trunk rings numbering 18–21 (typically 19–20); rings totaling 52–59 (typically 54–57); dorsal fin covering 1.5–6.5 (typically more than 2.5) trunk rings; dorsal fin with typically more than 32 rays.. 3

3a. Trunk rings numbering 19–21 (typically 20); dorsal fin covering 1.5–4.0 (typically less than 3.5) trunk rings; snout depth about 10% of snout length.............
....**chain pipefish** - *Syngnathus louisianae* (p. 147)

3b. Trunk rings numbering 18–21 (typically 19); dorsal fin covering 3.75–5.0 (typically 4.0 or more) trunk rings; snout depth about 20% of snout length..............
....**northern pipefish** - *Syngnathus fuscus* (p. 146)

Key to the Species of Searobins (Family Triglidae) in the Chesapeake Bay

1a. Second dorsal fin with 13 or 14 rays.. 2

1b. Second dorsal fin with 11 or 12 rays.. 3

2a. Chest completely scaled; opercular flap naked; branchiostegal rays black or dusky .... **northern searobin** - *Prionotus carolinus* (p. 150)

2b. Chest incompletely scaled; opercular flap scaled; branchiostegal rays pale ....
.... **leopard searobin** - *Prionotus scitulus* (p. 329)*

3a. Narrow black stripe extending along lateral line from head to caudal fin, with second, incomplete stripe ventral to it; pectoral fin with numerous narrowly separated dusky stripes................. **striped searobin** - *Prionotus evolans* (p. 151)

3b. Black stripes on body rudimentary or lacking; stripes on pectoral fin broad and widely spaced.........................**bighead searobin** - *Prionotus tribulus* (p. 329)*

Key to the Species of Butterfishes (Family Stromateidae) in the Chesapeake Bay

1a. Dorsal and anal fins (especially anal fin) greatly elongate anteriorly, with longest rays much longer than head length; pores absent near base of dorsal fin; body depth greater than 60% of SL.....................**harvestfish** - *Peprilus paru* (p. 152)

1b. Dorsal and anal fins slightly elongate anteriorly, with longest rays somewhat shorter than head length; series of well-developed conspicuous pores present near base of dorsal fin; body depth less than 60% of SL..
....**butterfish** - *Peprilus triacanthus* (p. 154)

Key to the Species of Jacks (Family Carangidae) in the Chesapeake Bay

1a. Body superficially naked, with scales minute and embedded..........................2
1b. Body scales easily observed and present over most of body..............................4

2a. Front of head rising gradually and forming smooth curve above eyes; pelvic fin of adults longer than upper jaw.......**African pompano** - *Alectis ciliaris* (p. 158)
2b. Front of head slightly concave in profile, rising nearly vertically then forming sharp angle above eye; pelvic fin of adults short, about one-fourth to one-third as long as upper jaw...3

3a. Anterior rays of second dorsal and anal fins notably elongate; body depth contained 2.3–2.8 times in FL; small juveniles with 4 or 5 faint interrupted bands on body..**lookdown** - *Selene vomer* (p. 170)
3b. Anterior rays of second dorsal and anal fins not notably elongate; body depth contained 1.8–2.3 times in FL; small juveniles with black oval spot on sides above straight part of lateral line..
....**Atlantic moonfish** - *Selene setapinnis* (p. 168)

4a. Posterior part of lateral line consisting of enlarged scutes (except scutes small and restricted to caudal peduncle in Atlantic bumper, *Chloroscombrus chrysurus*); pectoral fin long, more than 90% of head length.....................................5
4b. Posterior part of lateral line without enlarged scutes; pectoral fin short, less than 90% of head length..11

5a. Lower shoulder girdle margin bearing deep furrow with large papilla just above it (visible only when gill cover raised)...
....**bigeye scad** - *Selar crumenophthalmus* (p. 166)
5b. Lower shoulder girdle margin without deep furrow and adjoining papilla.......6

6a. Scutes covering entire length of lateral line to head (anterior scutes sometimes obscured by overgrowth of adjacent body scales)..
....**rough scad -** *Trachurus lathami* (p. 166)
6b. Scutes present only on posterior portion of lateral line......................................**7**

7a. Scutes very small and restricted to caudal peduncle; ventral profile more convex than dorsal profile; black blotch on upper part of caudal peduncle ....
.... **Atlantic bumper -** *Chloroscombrus chrysurus* (p. 164)
7b. Scutes prominent, extending anterior to caudal peduncle; ventral profile not more convex than dorsal profile; black blotch absent on caudal peduncle........**8**

8a. Anal fin with 16–18 branched rays; dorsal fin with 19–22 branched rays.........**9**
8b. Anal fin with 19–24 branched rays; dorsal fin with 22–28 branched rays **10**

9a. Dusky blotch on lower rays of pectoral fin; chest naked except for small patch of scales anterior to pelvic fin......................**crevalle jack -** *Caranx hippos* (p. 161)
9b. Dusky blotch absent on lower rays of pectoral fin; chest completely covered by scales ....**horse-eye jack -** *Caranx latus* (p. 162)

10a. Gill rakers on lower limb of first arch numbering 25–28; dorsal fin with 22–25 branched rays; anal fin with 19–21 branched rays; upper jaw terminating ventral to middle of eye..**blue runner -** *Caranx crysos* (p. 160)
10b. Gill rakers on lower limb of first arch numbering 18–21; dorsal fin with 25–28 branched rays; anal fin with 21–24 branched rays; upper jaw terminating ventral to anterior margin of eye .... **yellow jack -** *Caranx bartholomaei* (p. 162)

11a. Transverse groove dorsally and ventrally on caudal peduncle just anterior to cau-dal fin; anal-fin base decidedly shorter than second dorsal-fin base; body only slightly compressed **12**
11b. Transverse groove absent dorsally and ventrally on caudal peduncle just anterior to caudal fin; anal-fin base equal to or only slightly shorter than second dorsal-fin base; body strongly compressed **15**

12a. First dorsal fin with 8 spines **13**
12b. First dorsal fin with 7 spines... **14**

13a. Six dusky solid bars on body of juveniles (less than 30 cm, or 1 ft, FL), the third, fourth, and fifth extending into soft fin membranes; tips of caudal fin white .....
.... **banded rudderfish -** *Seriola zonata* (p. 174)
13b. Seven dusky irregular and broken bars present on body; third through seventh dusky bars extending into soft ray membranes of second dorsal fin and anal fin; eighth dusky bar small and situated at terminus of caudal peduncle; dusky rounded spot present on medial caudal-fin rays; caudal fin otherwise clear ....
.... **lesser amberjack -** *Seriola fasciata* (p. 330)*

14a. Longest ray in second dorsal fin contained about 7 times in FL ....
....**greater amberjack -** *Seriola dumerili* (p. 172)

14b. Longest ray in second dorsal fin contained about 5 times in FL ....
....**almaco jack -** *Seriola rivoliana* (p. 173)

15a. Snout pointed; body elongate, with greatest depth contained 3.4–3.9 times in
SL; posterior part of dorsal and anal fins with semidetached finlets; scales needle-like and partially embedded .... **leatherjack -** *Oligoplites saurus* (p. 165)
15b. Snout rounded; body short and deep, with greatest depth contained 1.3–2.6
times in SL; posterior part of dorsal and anal fins without semidetached finlets;
scales normal and oval-shaped.. **16**

16a. Dorsal fin with 22–27 rays; anal fin with 20–24 rays....................................
....**Florida pompano -** *Trachinotus carolinus* (p. 176)
16b. Dorsal fin with 17–21 rays; anal fin with 16–19 rays **17**

17a. Body without vertical bars; longest dorsal-fin rays not reaching beyond caudal-
fin base ....**permit -** *Trachinotus falcatus* (p. 178)
17b. Body usually with 2–5 (typically 4) narrow vertical bars that are black or silvery
in life; longest dorsal-fin rays reaching posteriorly well beyond caudal-fin base .
....**palometa -** *Trachinotus goodei* (p. 178)

Key to the Species of Remoras (Family Echeneidae) in the Chesapeake Bay

1a. Dorsal- and anal-fin elements numbering fewer than 30; body color nearly uniform, without lateral stripes; pectoral fin rounded 2
1b. Dorsal- and anal-fin elements typically numbering more than 30; body color pattern with dusky longitudinal stripe bounded above and below by whitish stripes;
pectoral fin pointed.. 3

2a. Pectoral-fin rays stiff.........................**marlinsucker -** *Remora osteochir* (p. 330)*
2b. Pectoral-fin rays soft and flexible ...... **whalesucker -** *Remora australis* (p. 330)*

3a. Sucking disk with 18–23 laminae (typically 21)...
.... **whitefin sharksucker -** *Echeneis neucratoides* (p. 330)*
3b. Sucking disk with 21–29 laminae (typically 23)...
....**sharksucker -** *Echeneis naucrates* (p. 181)

Key to the Species of Mackerels (Family Scombridae) in the Chesapeake Bay

1a. Dorsal fins widely separated; 2 small keels on either side of caudal peduncle; 5
dorsal and anal finlets present.. 2
1b. Dorsal fins close together; 2 small keels and large median keel between them
present on either side of caudal peduncle; 7–10 dorsal and anal finlets present
.... 3

2a. Dorsal-fin spines numbering 8–10; dorsum with numerous oblique lines that zigzag and undulate; dusky rounded blotches present on lower side and belly ... .**Atlantic chub mackerel** - *Scomber colias* (p. 184)

2b. Dorsal-fin spines numbering 11–13; dorsum with dark wavy oblique to vertical bars or streaks; dusky blotches absent on lower side and belly .... **Atlantic mackerel** - *Scomber scombrus* (p. 185)

3a. Teeth in jaws strong, compressed, and almost triangular or knife-like; body elongate, compressed **4**

3b. Teeth in jaws slender, conical, and hardly compressed; body fusiform, rounded (tuna-shaped)..**6**

4a. Lateral line abruptly curving downward below second dorsal fin; side without markings; first dorsal fin with 14–16 spines, anterior third of first dorsal fin without black pigment .... **king mackerel** - *Scomberomorus cavalla* (p. 186)

4b. Lateral line gradually curving downward to caudal peduncle; sides with brassy yellow spots; first dorsal fin with 17–19 spines; anterior third of first dorsal fin with black pigment **5**

5a. Side of body with spots only and without longitudinal stripes............................ .... **Spanish mackerel** - *Scomberomorus maculatus* (p. 188)

5b. Side of body with spots and 1 or 2 longitudinal stripes....................................... .... **cero** - *Scomberomorus regalis* (p. 333)*

6a. Dorsum and upper side with 7–12 oblique stripes extending below lateral line; dorsal surface of tongue without cartilaginous longitudinal ridges; 20–23 spines in first dorsal fin......................................**Atlantic bonito** - *Sarda sarda* (p. 183)

6b. Dorsum and upper side without oblique stripes; dorsal surface of tongue with 2 cartilaginous longitudinal ridges; 9–16 spines in first dorsal fin **7**

7a. Dorsum above lateral line with wavy dark reticulated lines; chest ventral to pectoral fin with 4 or 5 dusky spots; posterior part of body without scales ....**little tunny** - *Euthynnus alletteratus* (p. 182)

7b. Stripes, spots, or wavy lines absent from body; posterior part of body covered with very small scales....................... **bluefin tuna** - *Thunnus thynnus* (p. 333)*

Key to the Species of Barracuda (Family Sphyraenidae) in the Chesapeake Bay

1a. Scales in lateral line series typically numbering fewer than 115; origin of pelvic fin more anterior than origin of dorsal fin or tip of pectoral fin; fleshy tip absent on lower jaw...............................**guaguanche** - *Sphyraena guachancho* (p. 190)

1b. Scales in lateral line series typically numbering more than 120; origin of pelvic fin more posterior than origin of dorsal fin or tip of pectoral fin; fleshy tip present on lower jaw........................**northern sennet** - *Sphyraena borealis* (p. 190)

Key to the Species of Drums and Croakers (Family Sciaenidae) in the Chesapeake Bay

1a. Lower jaw with 1 or more barbels (sometimes minute and easily overlooked) 2
1b. Lower jaw without barbels 7

2a. Preopercular margin entire and without spines or bony "teeth"........................ .... **black drum** - *Pogonias cromis* (p. 206)
2b. Preopercular margin strongly to finely serrate and with spines or bony "teeth".. 3

3a. Lower jaw with row of minute barbels on each side; preopercular margin strongly serrate ....**Atlantic croaker** - *Micropogonias undulatus* (p. 204)
3b. Lower jaw with single thick barbel at tip; preopercular margin finely serrate 4

4a. Anal fin with 2 spines.......................... **sand drum** - *Umbrina coroides* (p. 212)
4b. Anal fin with 1 spine 5

5a. Scales on chest much smaller than scales on side above lateral line; posterior margin of pectoral fin pale .... **gulf kingfish** - *Menticirrhus littoralis* (p. 201)
5b. Scales on chest not much smaller than scales on side above lateral line; posterior margin of pectoral fin dusky or edged in black...6

6a. Anal fin typically with 8 soft rays (sometimes 7 or 9); longest spine of first dorsal fin extending beyond origin of second dorsal fin; side typically with blackish bars; nape with black bars forming V shape; pectoral fin with 18–21 rays.......... ....**northern kingfish** - *Menticirrhus saxatilis* (p. 202)
6b. Anal fin typically with 7 soft rays (rarely 6 or 8); spines of first dorsal fin not extending beyond origin of second dorsal fin; side with faint dusky bars; pectoral fin with 21–22 rays........**southern kingfish** - *Menticirrhus americanus* (p. 200)

7a. Upper jaw with pair of canine teeth..8
7b. Upper jaw without canine teeth 10

8a. Side silvery and without conspicuous spots ....**silver seatrout** - *Cynoscion nothus* (p. 195)
8b. Back and upper side with conspicuous black spots or blotches in rows............9

9a. Spots irregularly spaced; soft portion of dorsal fin without scales ....**spotted seatrout** - *Cynoscion nebulosus* (p. 194)
9b. Spots forming oblique streaks along scale rows; soft portion of dorsal fin with scales basally ....**weakfish** - *Cynoscion regalis* (p. 196)

10a. One or more black spots present dorsally on caudal peduncle; gill rakers on first arch numbering 12–14 ....**red drum** - *Sciaenops ocellatus* (p. 208)

10b. Spots absent at caudal-fin base; gill rakers on first arch numbering 20–36 .....**11**
11a. Dorsal fin with 29–35 branched rays; dusky spot on shoulder...........................
....**spot - *Leiostomus xanthurus*** (p. 198)
11b. Dorsal fin with 19–27 branched rays; dusky spot absent on shoulder............. **12**

12a. Preopercle entire or weakly serrate; mouth large and often very oblique; side marked with 7–9 dusky vertical bars ....
....**banded drum - *Larimus fasciatus*** (p. 210)
12b. Preopercle strongly serrate; mouth not large; side without dusky vertical bars ..
.... **13**

13a. Anal fin typically with 7–9 branched rays; skull cavernous and spongy-feeling to the touch... **star drum - *Stellifer lanceolatus*** (p. 210)
13b. Anal fin typically with 10 branched rays; skull not cavernous or noticeably spongy-feeling............................... **silver perch - *Bairdiella chrysoura*** (p. 192)

Key to the Species of Gobies (Family Gobiidae) in the Chesapeake Bay

1a. First dorsal fin with 6 flexible spines **2**
1b. First dorsal fin with 7 flexible spines **3**

2a. Teeth bicuspid and in males less than 60 mm (2.4 in) SL and in females; dark blotches separated by median pale area present at base of caudal fin in adults ...
....**lyre goby - *Evorthodus lyricus*** (p. 218)
2b. Teeth conical; large dark spot present above pectoral-fin base ....
....**darter goby - *Ctenogobius boleosoma*** (p. 218)

3a. Body largely scaled; single median interorbital pore present; second dorsal fin typically with 1 spine and 15–16 rays; caudal fin pointed.................................**4**
3b. Body scaleless or only 2 scales present on each side of caudal-fin base; 2 median interorbital pores present; second dorsal fin typically with 1 spine and 11–12 rays; caudal fin rounded **5**

4a. Body iridescent greenish blue with 4 or 5 tan to golden bars behind pectoral fin; yellow-green bands present on cheek; first dorsal fin with some red pigment....
....**green goby - *Microgobius thalassinus*** (p. 222)
4b. Body with numerous large dark blotches and no bright colors............................
.... **clown goby - *Microgobius gulosus*** (p. 332)*

5a. Body scaleless except for pair of large ctenoid scales on each side of caudal-fin base; second dorsal fin typically with 1 spine and 11 rays ....
....**seaboard goby - *Gobiosoma ginsburgi*** (p. 221)
5b. Body entirely scaleless; second dorsal fin typically with 1 spine and 12 rays.....**6**

6a. Short segment of lateral canal dorsal to opercle present; lateral midline without series of dark dots and dashes.................**naked goby - *Gobiosoma bosc*** (p. 220)

6b. Short segment of lateral canal dorsal to opercle absent; lateral midline with series of dark dots and dashes............ **code goby** - *Gobiosoma robustum* (p. 332)*

Key to the Species of Temperate Basses (Family Moronidae) in the Chesapeake Bay

1a. Body short, deep, and compressed; second spine of anal fin as long as third
spine; anal fin with 10 or fewer soft rays...
.... **white perch** - *Morone americana* (p. 223)
1b. Body elongate and stout; second spine of anal fin shorter than third spine; anal
fin with 10 or more soft rays................. **striped bass** - *Morone saxatilis* (p. 225)

Key to the Species of Perches (Family Percidae) in the Chesapeake Bay

1a. Dorsal fin with more than 10 spines; branchiostegal rays numbering 7 or 8; posterior margin of preopercle serrate.. 2
1b. Dorsal fin with 10 or fewer spines; branchiostegal rays numbering 5 or 6; posterior margin of preopercle smooth..
.... **tessellated darter** - *Etheostoma olmstedi* (p. 226)

2a. Body with 6–8 dusky crossbars; anal fin with 2 spines and 6–8 soft rays; jaws
without enlarged canine-like teeth; lower lobe of caudal fin without whitish tip
.... **yellow perch** - *Perca flavescens* (p. 228)
2b. Body without dark crossbars; anal fin with 2 spines and 12–13 soft rays; jaws
with numerous enlarged canine-like teeth; lower lobe of caudal fin with whitish
tip .... **walleye** - *Sander vitreus* (p. 330)*

Key to the Species of Combtooth Blennies (Family Blenniidae) in the Chesapeake Bay

1a. Caudal fin with 10 or 11 segmented rays; pectoral fin typically with 12 rays; dor-
sal fin typically with 11 spines...
....**striped blenny** - *Chasmodes bosquianus* (p. 230)
1b. Caudal fin typically with 13 segmented rays; pectoral fin typically with 13–15
rays; dorsal fin typically with 12 or 13 spines 2

2a. Enlarged canine teeth present posteriorly in jaws...
....**crested blenny** - *Hypleurochilus geminatus* (p. 332)*
2b. Enlarged canine teeth absent posteriorly in jaws ....
.... **feather blenny** - *Hypsoblennius hentz* (p. 231)

Key to the Species of Mojarras (Family Gerreidae) in the Chesapeake Bay

Morphometric characters apply only to specimens greater than 40 mm, or 1.6 in, SL.

1a. Margin of preopercle serrated; 10 or more gill rakers present on lower limb of first arch (including one at angle); anal fin with 8 rays (11 fin elements in total) ....**Irish pompano** - *Diapterus auratus* (p. 236)

1b. Margin of preopercle smooth; 9 or fewer gill rakers present on lower limb of first arch; anal fin with 7 rays (10 fin elements in total)....................................**2**

2a. Nine gill rakers present on lower limb of first arch; spinous dorsal fin distinctly tricolored, with jet-black blotch at tip, longitudinal white band (clear in preserved specimens) at middle of fin, and dusky pigment near base.................... ....**flagfin mojarra** - *Eucinostomus melanopterus* (p. 331)*

2b. Eight gill rakers present on lower limb of first arch; spinous dorsal fin often dusky to black at tip but not tricolored **3**

3a. Anal-fin base typically more than 16% of SL; groove formed on snout by ascending processes of premaxillary (i.e., premaxillary groove; readily seen by pulling upper jaw out and down) often constricted or crossed by scales, especially in larger specimens **4**

3b. Anal-fin base typically less than 16% of SL; premaxillary groove present but not constricted or crossed by scales..**5**

4a. Last spine of dorsal fin typically less than 7.5% of SL; pelvic fin typically less than 21.5% of SL; V- or U-shaped area on snout between nares, and longitudinal band posterior to this area (V and band) often unpigmented (especially in juveniles) ....**spotfin mojarra** - *Eucinostomus argenteus* (p. 236)

4b. Last spine of dorsal fin typically more than 7.5% of SL; pelvic fin typically more than 21.5% of SL; V and band on snout generally clearly discernible but pigmented...**silver jenny** - *Eucinostomus gula* (p. 236)

5a. Lateral line typically with 46 or more scales; least depth of caudal peduncle typically less than 10.5% of SL; pelvic fin typically less than 19% of SL; V and band on snout generally quite distinct and often nearly unpigmented in smaller specimens ....**slender mojarra** - *Eucinostomus jonesii* (p. 331)*

5b. Lateral line typically with 45 or fewer scales; least depth of caudal peduncle typically greater than 10.5% of SL; pelvic fin typically more than 19% of SL; V and band on snout generally darkly pigmented and often nearly obscured by pigment ....**tidewater mojarra** - *Eucinostomus harengulus* (p. 331)*

Key to the Species of Porgies (Family Sparidae) in the Chesapeake Bay

1a. Anterior teeth narrow, in close-set bands, and almost conical............................ ....**scup** - *Stenotomus chrysops* (p. 242)

1b. Anterior teeth very broad and incisor-like 2

2a. Large dusky "saddle" on caudal peduncle; first dorsal spine not procumbent (not directed forward) .... **spottail pinfish** - *Diplodus holbrookii* (p. 240)
2b. Dusky "saddle" on caudal peduncle absent; first dorsal spine procumbent (directed forward) 3

3a. Anterior teeth deeply notched; 2.5 rows of lateral molar-like teeth in each jaw; dusky blotch present near origin of lateral line, followed by 4–6 dusky vertical bars on body... **pinfish** - *Lagodon rhomboides* (p. 240)
3b. Anterior teeth with shallow notches or not notched; 3 rows of molar-like teeth present laterally in upper jaw and 2 rows in lower jaw; incomplete dark bar present across top of head, followed by 6 distinct vertical dark bars on body............
....**sheepshead** - *Archosargus probatocephalus* (p. 238)

Key to the Species of Snappers (Family Lutjanidae) in the Chesapeake Bay

1a. Tooth patch on roof of mouth V-shaped or crescentic, with posterior extension; prominent canine teeth in upper jaw much larger than canine teeth in lower jaw; dusky stripe present from snout through eye to upper margin of opercle in juveniles.. **gray snapper** - *Lutjanus griseus* (p. 243)
1b. Tooth patch on roof of mouth triangular, without posterior extension; canine teeth in both jaws very strong and equally well developed; stripe on snout through eye absent in fish at all sizes..
.... **cubera snapper** - *Lutjanus cyanopterus* (p. 330)*

Key to the Species of Grunts (Family Haemulidae) in the Chesapeake Bay

1a. Inside of mouth red; soft portions of dorsal and anal fins scaled nearly to outer margins... 2
1b. Inside of mouth not red; soft portions of dorsal and anal fins without scales......
.... **pigfish** - *Orthopristis chrysoptera* (p. 244)

2a. Dorsal fin with 12 spines .... **white grunt** - *Haemulon plumierii* (p. 331)*
2b. Dorsal fin with 13 spines .... **tomtate** - *Haemulon aurolineatum* (p. 331)*

Key to the Species of Sunfishes (Family Centrarchidae) in Chesapeake Bay waters

1a. Anal fin with 5–8 spines 2
1b. Anal fin typically with 3 spines.. 3

2a. First dorsal fin with 11–13 spines......... **flier** - *Centrarchus macropterus* (p. 246)
2b. First dorsal fin with 6–8 spines.. 8

3a. Caudal fin forked...**4**
3b. Caudal fin rounded...**9**

4a. Body elongate, its depth contained about 3 times in SL **5**
4b. Body short and deep, its depth contained less than 3 times in SL....................**6**

5a. Upper jaw terminating ventral or ventroposterior to posterior margin of orbit; typically 68 or fewer scales present in lateral series...
.... **largemouth bass -** *Micropterus salmoides* (p. 257)
5b. Upper jaw terminating ventral to middle of orbit; typically 68 or more scales present in lateral series .... **smallmouth bass -** *Micropterus dolomieu* (p. 256)

6a. Pectoral fin short and rounded; when bent forward, tip of pectoral fin usually not reaching front of eye ....**redbreast sunfish -** *Lepomis auritus* (p. 250)
6b. Pectoral fin long and sharply pointed; when bent forward, tip of pectoral fin reaching at least to front of eye..**7**

7a. Opercular flap flexible, blackish on posterior margin, and without red spot; posterior portion of second dorsal fin with dusky smudge or distinct black spot .....
....**bluegill -** *Lepomis macrochirus* (p. 254)
7b. Opercular flap stiff or moderately flexible, with orange or red spot on posterior tip; posterior portion of second dorsal fin without dusky smudge or distinct black spot.. **pumpkinseed -** *Lepomis gibbosus* (p. 252)

8a. Dorsal spines typically numbering 7 or 8; scattered spots present on side..........
.... **black crappie -** *Pomoxis nigromaculatus* (p. 258)
8b. Dorsal spines typically numbering 6; dusky markings present on side, forming rough vertical bars....................... **white crappie -** *Pomoxis annularis* (p. 329)*

9a. Spinous portion of dorsal fin slightly greater in height than soft-rayed portion…....................... **blackbanded sunfish -** *Enneacanthus chaetodon* (p. 248)
9b. Spinous portion of dorsal fin lower than or equal in height to soft-rayed portion
....**10**

10a. Dusky bars present on side ...... **banded sunfish -** *Enneacanthus obesus* (p. 248)
10b. Dusky bars absent on side ....
.... **bluespotted sunfish -** *Enneacanthus gloriosus* (p. 248)

Key to the Species of Wrasses (Family Labridae) in the Chesapeake Bay

1a. Snout pointed; scales large, numbering about 40 in lateral series; cheek and opercle with scales .... **cunner -** *Tautogolabrus adspersus* (p. 262)
1b. Snout blunt; scales small, numbering about 70 in lateral series; cheek and opercle with few or no scales.................................... **tautog -** *Tautoga onitis* (p. 260)

Key to the Species of Sea Basses and Groupers (Family Serranidae) in the Chesapeake Bay

1a. Caudal fin with 3 distinct lobes; dorsal-fin spines with fleshy tabs or filaments **2**

1b. Caudal fin truncate or lunate but never with 3 distinct lobes; dorsal-fin spines without fleshy tabs or filaments.. **3**

2a. Predominant color pattern dark brown to bluish black, with light spots forming longitudinal stripes; juveniles with broad dark lateral stripe and dark blotch at base of spinous dorsal fin.............. **black sea bass - *Centropristis striata*** (p. 264)

2b. Predominant color pattern olive brown or bronze, with dark blotches forming vertical bars; dark black blotch on middle of dorsal-fin base .... **rock sea bass - *Centropristis philadelphica*** (p. 329)*

3a. Anal fin with 3 spines and 8 soft rays; caudal fin rounded; dorsal spines short, with membranes between them deeply notched; predominant color pattern of gray or greenish ground color with small dark spots scattered on upper part of head and body.........................**Goliath grouper - *Epinephelus itajara*** (p. 329)*

3b. Anal fin with 3 spines and 11 soft rays; caudal fin emarginate; dorsal spines not noticeably short, with membranes between them not deeply notched; predominant color pattern of brownish gray ground with dark wormlike markings on sides... **gag - *Mycteroperca microlepis*** (p. 266)

Key to the Species of Filefishes (Family Monacanthidae) in the Chesapeake Bay

1a. Pelvic spine prominent, in form of barbed process in midline at end of pelvis ... .**planehead filefish - *Stephanolepis hispidus*** (p. 272)

1b. Pelvic spine absent **2**

2a. Head and body olive with blue or green reticulation and numerous black or maroon spots; dorsal fin with 43 or more rays; anal fin with 46 or more rays .... **scrawled filefish - *Aluterus scriptus*** (p. 270)

2b. Head and body grayish to brownish with large irregular whitish blotches and numerous small orange yellow spots; dorsal fin with 39 or fewer rays; anal fin with 41 or fewer rays .... **orange filefish - *Aluterus schoepfii*** (p. 270)

Key to the Species of Puffers (Family Tetraodontidae) in the Chesapeake Bay

1a. Dorsal fin with 13 or more rays; anal fin usually with 12 or more rays; caudal-fin margin distinctly concave or slightly crescent shaped; scales (prickles) restricted to belly ....**smooth puffer - *Lagocephalus laevigatus*** (p. 274)

1b. Dorsal fin with 9 or fewer rays; anal fin with 8 or fewer rays; caudal fin margin slightly rounded or truncate; prickles present over most of body, more pronounced on dorsum 2

2a. Back grayish green with ill-defined black spots and blotches; sides yellowish with several irregular bars; intense black spot present in pectoral-fin axil.......... ....**northern puffer** - *Sphoeroides maculatus* (p. 275)

2b. Back brown to black, with pale lines forming reticulated pattern that approximates concentric circles from dorsal view; numerous dusky spots on side and back; intense black spot absent in pectoral-fin axil .... **checkered puffer** - *Sphoeroides testudineus* (p. 333)*

Key to the Species of Righteye Flounders (Family Pleuronectidae) in the Chesapeake Bay

1a. Dorsal fin with 92 or more rays; caudal-fin margin concave ....**Atlantic halibut** - *Hippoglossus hippoglossus* (p. 333)*

1b. Dorsal fin with 91 or fewer rays; caudal-fin margin convex............................... 2

2a. Dorsal fin with 73–91 rays; lateral line distinctly arched over pectoral fin .... **yellowtail flounder** - *Limanda ferruginea* (p. 333)*

2b. Dorsal fin with 60–76 rays; lateral line nearly straight anteriorly ....**winter flounder** - *Pseudopleuronectes americanus* (p. 287)

Key to the Species of Sand Flounders (Family Paralichthyidae) in the Chesapeake Bay

1a. Lateral line strongly arched above pectoral fin... 2

1b. Lateral line nearly straight throughout its length... 4

2a. Ocular side with prominent ocellated spots.. 3

2b. Ocular side without prominent ocellated spots.. .... **southern flounder** - *Paralichthys lethostigma* (p. 284)

3a. Interorbital distance much less than diameter of pupil; ocular side with 4 prominent ocellated black spots .... **fourspot flounder** - *Paralichthys oblongus* (p. 286)

3b. Interorbital distance about equal to diameter of pupil; ocular side with triangular pattern of ocelli, consisting of 1 ocellus on midline and 2 ocelli on body edges at caudal peduncle ....**summer flounder** - *Paralichthys dentatus* (p. 283)

4a. Mouth small; maxilla about 25% of HL, terminating posteriorly near anterior edge of lower eye; jaws on blind side arched; front teeth in both jaws same size as posterior teeth.. 5

4b. Mouth moderately large; maxilla about 35% of HL, extending to middle of lower eye; jaws on blind side not arched; front teeth in jaws larger than posterior teeth .... **bay whiff** - *Citharichthys spilopterus* (p. 280)

5a. Body elongate, depth about 43–51% of SL; primary body scales with overlapping secondary scales; gill rakers numbering about 13 on lower limb of first arch .....
.... **smallmouth flounder** - *Etropus microstomus* (p. 282)

5b. Body ovate, with depth about 50–58% of SL; primary body scales without overlapping secondary scales; gill rakers numbering 6–9 on lower limb of first arch
.... **fringed flounder** - *Etropus crossotus* (p. 280)

APPENDIX 4. Fish Species Rarely Recorded from the Chesapeake Bay

The fish species listed here have been recorded from the Chesapeake Bay, but in many instances their occurrence has been reported only once. A specimen that validates this occurrence may or may not exist; the Virginia Institute of Marine Science (VIMS) holds many of the voucher (validating) specimens. Species listed in this appendix are not expected to be found in the bay in any abundance or with any regularity.

Nurse shark - *Ginglymostoma cirratum* (Bonnaterre, 1788)
The nurse shark was reported from the "southern part of Chesapeake Bay" in 1877; it has not been recorded from the bay since that time.

Lemon shark - *Negaprion brevirostris* (Poey, 1868)
Lemon sharks occur occasionally in coastal lagoons along the Eastern Shore as well as along Atlantic beaches. There is a single VIMS record (uncatalogued, specimen released) of a lemon shark collected at the bay mouth near Fishermans Island in June 1981.

Tiger shark - *Galeocerdo cuvier* (Péron and Lesueur, 1822)
Tiger sharks are rare visitors to the lower Chesapeake Bay, although they are common in coastal waters in summer. A single record (VIMS 7382) is known from Smith Island shoal at the bay mouth.

Blacktip shark - *Carcharhinus limbatus* (Müller and Henle, 1839)
Blacktip sharks are common in coastal lagoons along the Eastern Shore as well as along Atlantic beaches. They have been occasionally taken at the bay mouth near Fishermans Island, and there is a single VIMS record (uncatalogued, specimen released) of a blacktip shark collected there in June 1980.

Atlantic torpedo - *Torpedo nobiliana* Bonaparte, 1835
Known from both tropical and temperate waters, Atlantic torpedoes are rare visitors to the lower Chesapeake Bay. The last recorded collection was from 1922. Atlantic torpedoes can reach almost 2 m (6.6 ft) in length and attain a weight of 90 kg (198 lb). A large Atlantic torpedo is capable of emitting electric shocks that are sufficient to immobilize a careless human temporarily.

Spotted eagle ray - *Aetobatus narinari* (Euphrasen, 1790)
This circumtropical species has been reported in the Chesapeake Bay only as a rare summertime visitor, most recently in 1956. This coastal species is often observed at or near the surface.

Manta - *Manta birostris* (Walbaum, 1792)
Manta are rare visitors to the lower Chesapeake Bay during the summer months but are common offshore. Mantas are usually observed swimming or basking at the water's surface but may also rest on the bottom.

Bonefish - *Albula vulpes* (Linnaeus, 1758)
Bonefish are known from both the eastern and western Atlantic. Bonefish are rare visitors during the late summer to the lower Chesapeake Bay. They frequent shallow inshore waters, including bays and estuaries. Bonefish sometimes feed in water so shallow that their dorsal and caudal fins break the surface.

Speckled worm eel - *Myrophis punctatus* Lütken, 1852
Although speckled worm eels are common in the lower Chesapeake Bay in the larval stage, adults are rare visitors to the lower bay during the late summer and fall. The species is usually found in brackish waters of tidal creeks and protected bays but can be found in depths to at least 20 m (65 ft).

Hardhead catfish - *Ariopsis felis* (Linnaeus, 1766)
Hardhead catfish are rare summer visitors to the lower Chesapeake Bay and reach as far as the York River. The species is found in shallow turbid waters over muddy bottoms.

Rainbow trout - *Oncorhynchus mykiss* (Walbaum, 1792)
During the mid-1800s, rainbow trout were introduced into some Maryland tributaries of the bay. As rainbow trout are intolerant of water temperatures greater than 21°C (70°F), annual stocking is necessary to maintain the population.

Coho salmon - *Oncorhynchus kisutch* (Walbaum, 1792)
A single record of a coho salmon from the Chesapeake Bay is known. The specimen (VIMS 2473) was collected near the mouth of the York River; its fins had been clipped and it had apparently been released as a smolt in North River, Massachusetts, the previous year.

Atlantic salmon - *Salmo salar* Linnaeus, 1758
Atlantic salmon are found on both sides of the Atlantic Ocean. Spawning runs probably do not occur south of the Connecticut River, but individuals stray southward. Only a few records of Atlantic salmon in the Chesapeake Bay are known, and these are from near the bay mouth.

Brook trout - *Salvelinus fontinalis* (Mitchill, 1814)
Brook trout are native to the mountain headwater streams of the Chesapeake Bay region. They are intolerant of water temperatures greater than 24°C (75°F) and, consequently, are maintained on the Chesapeake coastal plain only by restocking. Sea-run populations of brook trout are known as far southward as Long Island, and individuals have been collected in Chesapeake Bay waters of 5‰ or greater.

Brown trout - *Salmo trutta* Linnaeus, 1758
Brown trout have been stocked in both streams and lakes in the Chesapeake Bay region, and they sometimes penetrate into waters of 5‰ or greater. Water temperatures greater than 27°C (81°F) are lethal to brown trout, so the species can be maintained in coastal streams of the Chesapeake Bay region only by annual restocking.

Snakefish - *Trachinocephalus myops* (Forster, 1801)
A single record (VIMS 11352) of this species exists for the Chesapeake Bay. An 83-mm (3.3-in) SL snakefish was collected in September 2005 on the bay side of the Chesapeake Bay Bridge-Tunnel.

White hake - *Urophycis tenuis* (Mitchill, 1814)
White hake are rare springtime visitors to the lower Chesapeake Bay. They are typically associated with soft muddy bottoms of the continental shelf and upper slope. As bay waters warm in late spring, this species migrates to deeper water offshore.

Sargassumfish - *Histrio histrio* (Linnaeus, 1758)
No specimens of sargassumfish collected in the Chesapeake Bay are known to the present authors. From an 1876 publication, the sargassumfish reportedly "occurs in the oyster regions of the Chesapeake Bay, but is perhaps quite uncommon." Although the sargassumfish is primarily a tropical species, it is conceivable that it could be carried northward with *Sargassum* seaweed and drift or be blown into the bay.

Longnose batfish - *Ogcocephalus corniger* Bradbury, 1980
Longnose batfish were reported as rare in the lower Chesapeake Bay during the late 1800s. No recent records of longnose batfish are known from the bay.

Atlantic agujon - *Tylosurus acus* (Lacepède, 1803)
Agujon are primarily an offshore species, but they are occasional late spring and summer visitors to the Chesapeake Bay, extending as far up the bay as the lower Susquehanna River.

Atlantic flyingfish - *Cheilopogon melanurus* (Valenciennes, 1847)
In 1876, the Atlantic flyingfish was reported (dubiously) to occur as far north in the Chesapeake Bay as the mouth of the Potomac River. Recent records of Atlantic flyingfish from Chesapeake Bay waters are not known. This surface-dwelling coast-

al species occasionally enters bays and may be a rare summer visitor to the lower Chesapeake Bay.

Squirrelfish - *Holocentrus adscensionis* (Osbeck, 1765)
Squirrelfish range from Virginia to Florida, throughout the Gulf of Mexico and Caribbean, to Brazil. They are rare visitors to the lower bay in late summer.

Bluespotted cornetfish - *Fistularia tabacaria* Linnaeus, 1758
Bluespotted cornetfish are rare to occasional visitors to the lower bay in late summer and early autumn. Within the bay, they frequent seagrass beds, but in tropical portions of their range they are often found on coral reefs.

Leopard searobin - *Prionotus scitulus* Jordan and Gilbert, 1882
Leopard searobins are a warm-water species that has been recorded from the Virginia portion of the Chesapeake Bay, extending as far north in the bay as the Potomac River.

Bighead searobin - *Prionotus tribulus* Cuvier, 1829
This species is a rare late-summer visitor to the lower Chesapeake Bay, recorded from the York River and near the bay mouth.

Sea raven - *Hemitripterus americanus* (Gmelin, 1789)
Sea ravens inhabit cold-temperate waters and are rare visitors to the lower Chesapeake Bay. They are not abundant south of New Jersey and may be expected to occur in the Chesapeake Bay region during the winter and early spring.

Flying gurnard - *Dactylopterus volitans* (Linnaeus, 1758)
Flying gurnards are rare visitors to the lower Chesapeake Bay in the late summer and fall.

Goliath grouper - *Epinephelus itajara* (Lichtenstein, 1822)
In the Atlantic, Goliath groupers are known from Virginia to Brazil, including the Gulf of Mexico, Caribbean Sea, and Bermuda. They inhabit reefs, ledges, and wrecks offshore and bridge pilings, jetties, and submerged mangrove roots inshore.

Rock sea bass - *Centropristis philadelphica* (Linnaeus, 1758)
A single specimen (VIMS 11979) of rock sea bass measuring 21 cm (8.3 in) TL was collected on the bay side of the first tunnel of the Chesapeake Bay Bridge-Tunnel in November 2007. This is the first record of the rock sea bass in Chesapeake Bay waters.

White crappie - *Pomoxis annularis* Rafinesque, 1818
This introduced freshwater species occurs in lakes, ponds, and slow-moving streams

of the Chesapeake Bay region. Although tolerant of very low-salinity waters, white crappies rarely penetrate into waters of 5‰ or greater.

Walleye - *Sander vitreus* (Mitchill, 1818)
Walleyes are recorded from various Western Shore tributaries of the Chesapeake Bay. Reported to enter brackish waters, walleyes rarely penetrate waters of 5‰ or greater.

Bigeye - *Priacanthus arenatus* Cuvier, 1829
Like short bigeyes, this species is a rare to occasional visitor to the lower Chesapeake Bay during the summer. Although juvenile and larval bigeyes are commonly recorded as far north as Nova Scotia, no adults are known from north of North Carolina, and thus juveniles probably do not survive the winter in more northerly latitudes.

Short bigeye - *Pristigenys alta* (Gill, 1862)
Short bigeyes are rare to occasional visitors to the lower Chesapeake Bay during the summer. Juvenile and larval fish occur in drifting weed lines near the surface off-shore and are commonly recorded as far north as Maine. However, no large adults are known from north of Cape Hatteras, and the juveniles probably do not survive the winter in those latitudes.

Whitefin sharksucker - *Echeneis neucratoides* Zuiew, 1789
Whitefin sharksuckers are rare summer visitors to the lower Chesapeake Bay, reaching as far north as the mouth of the York River. This species is far less common than the sharksucker (*E. naucrates*) and there is little published information on this fish.

Whalesucker - *Remora australis* (Bennett, 1840)
Whalesuckers are rare summer visitors to the lower Chesapeake Bay. They have been found attached to marine mammals.

Marlinsucker - *Remora osteochir* (Cuvier, 1829)
Marlinsuckers are rare summer visitors to the lower Chesapeake Bay, reaching as far north as the York River. Although occasionally free-living, this species is usually found attached to billfishes, especially white marlin and sailfish.

Dolphinfish - *Coryphaena hippurus* Linnaeus, 1758
Although adult dolphinfish are not known (and not expected) from the Chesapeake Bay, small juveniles (<7 cm, or <3 in, SL) have been collected in the lower bay.

Lesser amberjack - *Seriola fasciata* (Bloch, 1793)
Lesser amberjacks are rare visitors to the lower Chesapeake Bay in late summer.

Cubera snapper - *Lutjanus cyanopterus* (Cuvier, 1828)
Cubera snappers are known from the Chesapeake Bay to Brazil, including the Gulf

of Mexico and the Caribbean Sea. This tropical species is rare north of Florida. Adult cubera snappers are rare visitors to the lower bay during autumn.

Atlantic tripletail - *Lobotes surinamensis* (Bloch, 1790)
Atlantic tripletails are occasional visitors to the lower Chesapeake Bay during the summer and autumn. The largest tripletail known from the Chesapeake Bay was a 79-cm (30-in) fish (VIMS 7779) collected from a pound net at the mouth of the James River off Buckroe Beach.

Tidewater mojarra - *Eucinostomus harengulus* Goode and Bean, 1879
A single specimen (VIMS 0759) is known from Chesapeake Bay waters. This individual, of 63 mm (2.5 in) SL, was collected at the mouth of the York River off Gloucester Point in 1948.

Slender mojarra - *Eucinostomus jonesii* (Günther, 1879)
Slender mojarras are rare visitors to the Chesapeake Bay in summer and fall.

Flagfin mojarra - *Eucinostomus melanopterus* (Bleeker, 1863)
Primarily a tropical and subtropical species, flagfin mojarras are rare visitors to the Chesapeake Bay in the summer and fall.

Tomtate - *Haemulon aurolineatum* Cuvier, 1830
Known from the Chesapeake Bay from a record in 1876, the tomtate has not been recorded in the bay since that time.

White grunt - *Haemulon plumierii* (Lacepède, 1801)
Known from the mouth of the Potomac River from a record in 1876, the white grunt has not been recorded in the bay since that time.

Atlantic threadfin - *Polydactylus octonemus* (Girard, 1858)
Rare visitors to the lower Chesapeake Bay in late summer, Atlantic threadfin occur on sand bottoms near beaches, often in water only a few centimeters deep.

Barbu - *Polydactylus virginicus* (Linnaeus, 1758)
Barbu are rare tropical strays recorded from the lower Chesapeake Bay. Their ecology is similar to that of Atlantic threadfin.

Red goatfish - *Mullus auratus* Jordan and Gilbert, 1882
Although recorded from as far north as Cape Cod, red goatfish are primarily a tropical-subtropical species. They are summer to early fall visitors to the lower bay, and only juveniles (40–43 mm, or 1.6–1.7 in, SL) have been collected (VIMS 11347).

Dwarf goatfish - *Upeneus parvus* Poey, 1852
Dwarf goatfish are rare summer visitors to the Chesapeake Bay that have been collected as far north as Mobjack Bay near the York River mouth.

Yellow chub - *Kyphosus incisor* (Cuvier, 1831)
A yellow chub (VIMS 0017) was collected on the lower Eastern Shore off Kiptopeke, Virginia. This individual, the only adult yellow sea chub recorded north of the Bahamas, was a female measuring 36 cm (17 in) SL. Juvenile yellow chubs are expected to enter the lower bay as rare summer stragglers.

Bermuda chub - *Kyphosus sectatrix* (Linnaeus, 1758)
A juvenile Bermuda chub was collected at Fishermans Island near the Chesapeake Bay Bridge-Tunnel. Other Bermuda chubs are expected to enter the lower bay as rare summer stragglers.

Blue parrotfish - *Scarus coeruleus* (Edwards, 1771)
A single record of the blue parrotfish from Chesapeake Bay waters is known and is based on an 1894 collection by pound net from the Potomac River, off St. George Island.

Crested blenny - *Hypleurochilus geminatus* (Wood, 1825)
Only a few specimens of crested blennies are known from the Chesapeake Bay (VIMS 9086 and 11776). All were collected near the bay mouth in either September or November.

Fat sleeper - *Dormitator maculatus* (Bloch, 1792)
A single record of the fat sleeper is known from the Chesapeake Bay: an individual of 21 mm (0.8 in) SL was collected from a small creek near the Nanticoke River in October 1990 by electrofishing in less than 1 m (3.3 ft) of water.

Clown goby - *Microgobius gulosus* (Girard, 1858)
Clown gobies have been recorded only once from within the Chesapeake Bay. That record is of 32 specimens collected in a single trawl tow at the mouth of the Patuxent River in February 1962. This record is considered doubtful and perhaps the result of a collection label error, given that the typical range of this species is from St. Johns River, Florida, to Corpus Christi, Texas.

Code goby - *Gobiosoma robustum* Ginsburg, 1933
Ten specimens of code goby were reported as having been collected in the same tow with the clown gobies at the mouth of the Patuxent River in February 1962 (see preceding species account). Like the clown goby record, this record is considered doubtful, given the considerable northerly extension from code gobies' typical range of Cape Canaveral to Yucatán.

Cero - *Scomberomorus regalis* (Bloch, 1793)
This species was once recorded from "Chesapeake Bay near the ocean" in the 1870s. Cero are uncommon north of Florida.

Bluefin tuna - *Thunnus thynnus* (Linnaeus, 1758)
The bluefin tuna, an oceanic species, is reported to have been collected once in the twentieth century (in 1909) by pound net near the Chesapeake Bay mouth. This epipelagic species is commonly taken in June in Virginia coastal waters.

Atlantic halibut - *Hippoglossus hippoglossus* (Linnaeus, 1758)
A cold-temperate species, the Atlantic halibut is only occasionally encountered south of New York and has been recorded once from within the Chesapeake Bay. That record is an individual of 1.8 m (6 ft) TL, collected in a pound net near the Potomac River mouth.

Yellowtail flounder - *Limanda ferruginea* (Storer, 1839)
Known from the Chesapeake Bay from an 1876 record, the yellowtail flounder has not been recorded in the bay since that time. Yellowtail flounder are only rarely encountered south of New York and are unlikely to occur in the Chesapeake Bay, except during winter months.

Spotted whiff - *Citharichthys macrops* Dresel, 1885
A single specimen (VIMS 11345) of spotted whiff measuring 10.5 cm (4.1 in) TL was collected on the bay side of the Chesapeake Bay Bridge-Tunnel in September 2004.

Scrawled cowfish - *Acanthostracion quadricornis* (Linnaeus, 1758)
A single record, dating from 1877, exists for scrawled cowfish from the Chesapeake Bay.

Checkered puffer - *Sphoeroides testudineus* (Linnaeus, 1758)
A single record, dating from 1877, exists for checkered puffer from the Chesapeake Bay.

Porcupinefish - *Diodon hystrix* Linnaeus, 1758
A single record, dating from 1876, exists for porcupinefish from the Chesapeake Bay and is based on a specimen from the Potomac River along St. Mary's County, Maryland.

Ocean sunfish - *Mola mola* (Linnaeus, 1758)
A single record of ocean sunfish is known from Chesapeake Bay waters. This specimen (VIMS 8120) was foundering off a beach near the mouth of the Potomac River.

GLOSSARY OF SELECTED
TECHNICAL TERMS

Abdominal: That which pertains to the belly; often used to refer to the ventral portion of the body from the thorax to the anal opening.

Adipose eyelid: A fatty, transparent tissue covering the eye in some fishes.

Adipose fin: A fleshy, fin-like, rayless structure situated on the dorsal ridge between the dorsal and tail fins, which in some species is fused to the tail and separated from it only by a slight notch.

Aestivation: Inactivity associated with very dry environmental conditions.

Anal fin: The vertical fin on the ventral side of fishes just anterior to the tail and posterior to the anus.

Axil: The armpit or backside of the pectoral- or pelvic-fin base.

Axillary scale: A modified, usually elongate, scale attached to the upper or anterior base of the pectoral or pelvic fins in certain fishes.

Barbel: A fleshy, tactile, enlarged flap or icicle-shaped projection, usually situated about the lips, chin or nose; varies considerably in number and size in the various species of fishes.

Caudal fin: The tail fin.

Caudal peduncle: That region of the body between the base of the posterior ray of the anal fin and the base of the tail fin.

Clasper: Modified pelvic fin used for reproduction in male sharks, rays, and skates.

Compressed: Flattened laterally.

Concave: Arched inward; a concave fin is one in which the central soft rays or spines are shorter than the anterior and posterior soft rays or spines, thereby causing an inwardly curved distal edge to the fin.

Convex: Arched or rounded outward; a convex fin is one in which the central soft rays or spines are longer than the anterior and posterior soft rays or spines, thereby causing an outwardly curved distal edge to the fin.

Ctenoid scales: Having a comb-like margin of tiny prickles (ctenii) on the exposed or posterior field; the ctenii cause the scales to feel rough to the touch when stroked.

Cycloid scales: Smooth-edged scales having an evenly curved posterior border and no trace of minute spines or ctenii.

Depressed: Flattened dorsoventrally.

Dorsal: The back or upper part of the body.

Dorsal fins: The fins of the midline of the back; usually two, a spiny-rayed dorsal fin followed by a soft-rayed dorsal fin that may or may not be connected.

Esca: A fleshy bulb-like structure located at the distal end of the angling apparatus (illicium) characteristic of goosefishes, batfishes, and frogfishes.

Euryhaline: Showing a broad tolerance to a wide range of salinity.

Falcate: Curved like a sickle; a fin is falcate when it is deeply concave, having the middle rays much shorter than are the anterior and posterior rays.

Fecundity: The reproductive potential of a female, typically based on the number of eggs produced.

Fimbriate: Fringed at the margin.

Finlet: Separated parts of divided dorsal and anal fins.

Frenum: A connecting membrane that binds a part or parts together, such as the skin that joins the upper jaw to the snout.

Ganoid or ganoid scales: Diamond- or rhombus-shaped scales consisting of bone covered with superficial enamel.

Genital papilla: A small blunt fleshy projection behind the anal opening; present in gobies and other fishes.

Gill: A respiratory organ in aquatic animals consisting chiefly of filamentous outgrowths for breathing oxygen dissolved in water.

Gill arch: The branchial skeleton that contains the gill rakers and the gill lamellae.

Gill raker: Projections on the inner edge of gill arches that prevent food particles from passing outward through the gill slits.

Groundfishes: Fishes that live on or near the bottom of the water body they inhabit.

Gular plate: A large, bony plate between ventral portions of the arms of the lower jaw.

Heterocercal: A caudal fin is said to be heterocercal when the posterior end of the vertebral column (backbone) is flexed upward, entering and continuing near the end of the upper or dorsal lobe of the caudal fin (but does not enter the filament if one is present), that lobe being better developed and often longer than the lower lobe.

Homocercal: A caudal fin is said to be homocercal when the posterior end of the vertebral column does not flex upward and does not enter either lobe of the caudal fin but ends in a hypural plate, the two lobes of the tail being nearly equal or equal.

Illicium: The angling apparatus or "fishing pole" of goosefishes, batfishes, and frogfishes. This structure represents the remains of the spinous dorsal fin.

Inferior mouth: A mouth is inferior when located near to or on the ventral side of the head and when the snout overhangs the upper lip.

Inner narial groove: An external groove along the front margin of the head of some hammerhead sharks extending from the narial opening back toward the median part of the head.

Interradial membranes: Membranes between fin rays or spines.

Isthmus: The narrow portion of the breast lying between the gill chambers and separating them.

Jugular: Pertaining to the throat; the pelvic fins are "jugular" when located anterior to the pectoral fins.

Lanceolate: Pointed; lance-shaped.

Lateral line: A line formed by a series of sensory tubes and pores, extending backward from the head along the sides of the body. The lateral line is complete when all pores are present and the line reaches to the base of the caudal fin; incomplete when the line does not extend as far as the base of the caudal fin; and absent when no tubes or pores are present.

Mandible: The lower jaw.

Metric ton (mt): Equivalent to 1,000 kilograms (kg) or 2,205 pounds (lb).

Nape: That small area on the dorsal surface of a fish beginning immediately posterior to the back of the head.

Opercle or operculum: The large, very flat, thin bones on each side of the head of fishes, that cover the gills; also called gill cover.

Oviparous: Producing external eggs.

Paired fins: The pectoral and pelvic fins.

Papilla: A small, fleshy projection.

Pectoral fins: Paired fins attached to the shoulder on the side of the body.

Pelvic fins: Paired fins on the ventral side of the body; sometimes called ventral fins. They may be posterior to the pectorals (abdominal), ventral to the pectorals (thoracic), or anterior to the pectorals (jugular).

Precaudal pit: A depression or indentation on the tail of a shark located at the dorsal and/or ventral midline just anterior to the caudal fin.

Preopercle or preoperculum: The bone lying anterior to the opercle and comprising the forepart of the gill cover.

Ray: Any of the soft or hard rays of the fins or any of the spines. A *soft ray* is usually flexible, branched, bilaterally paired, and segmented; it may be either a principal or a rudimentary ray. A *hard ray* is a hardened soft ray that may be a simple spinous ray or the consolidated product of branching, as in the catfishes. A *true spinous ray* is an unpaired structure that is without segmentation, sharpened apically, and usually stiff.

Scute: A modified scale, often spiny or keeled. Scutes are found along the ventral midline of some species and along the lateral line of others.

Serrated: Notched or toothed, something like a saw.

Spiracle: An opening dorsoposterior to the eye in rays and some sharks.

Stenohaline: Tolerant of a narrow range of salinity.

Terminal mouth: When the upper and lower jaw form the extreme anterior tip of the head.

Thoracic: In the region of the thorax or chest; pelvic fins are thoracic when inserted anteroventrally to the pectoral fins, instead of considerably posterior to the pectoral fins, as is more often the case.

Venter: That portion of the body relating to the belly.

Vertical fins: The dorsal, anal, and caudal fins, which are unpaired and located on the median line of the body, in contrast to the pectoral and pelvic fins, which are paired.

Viviparous: Producing live young.

INDEX TO SCIENTIFIC NAMES

INDEX TO COMMON NAMES

Photo by David Weisberg

EDWARD O. MURDY is an ichthyologist who has conducted research on fishes in several countries. He has served in various capacities within the National Science Foundation, where he is currently Director of the National Science Foundation office in Tokyo, Japan. He is an adjunct professor at George Washington University, where he teaches marine biology. Murdy received a B.S. and an M.S. in Biology from Old Dominion University and a Ph.D. in Wildlife & Fisheries Sciences from Texas A&M University. He has published dozens of scientific papers and is the author of two other books.

Photo by John Carlson

JOHN A. (JACK) MUSICK is the Marshall Acuff Professor Emeritus in Marine Science at the Virginia Institute of Marine Science (VIMS), College of William and Mary. He earned his B.A. in Biology from Rutgers University in 1962 and his M.A. and Ph.D. in Biology from Harvard University in 1964 and 1969, respectively. Dr. Musick has been awarded the Excellence in Fisheries Education Award by the American Fisheries Society. He has published more than 150 scientific papers and coauthored or edited 21 books focused on ecology and conservation. He is an elected Fellow of the American Association for the Advancement of Science.

Photo by Andrew Middleditch

VAL KELLS is a Marine Science Illustrator who works with curators, educators, designers, authors, and publishers. She creates a wide variety of accurate and aesthetic illustrations for educational and interpretive exhibits in public aquariums, museums, and nature centers. Her artwork has also been published in numerous books and periodicals. She most recently illustrated and coauthored *A Field Guide to Coastal Fishes: From Maine to Texas*, a five-year project. Val completed her formal training in Scientific Illustration and Communication at the University of California, Santa Cruz. Val is an avid fisherman and naturalist who lives with her family in Virginia.